More praise for

THE NATURAL HISTORY OF THE RICH

"This book . . . may change forever our perception of the urge to make money."
— *Financial Times*

"As one who has long had a fascination with the idiosyncrasies of the possessors of great wealth, I was intrigued [by] *The Natural History of the Rich*. It is anecdotal, witty, and wonderfully informative."
—Dominick Dunne, author of
The Two Mrs. Grenvilles and *An Inconvenient Woman*

"Hilarious."
—*New York*

"[A] literate, gossipy, and altogether engaging romp."
—*Smithsonian*

"Amusing and insightful."
—*Boston Herald*

"A witty compendium of gossip, anecdotes, history and sociobiological research."
—*Town and Country*

"In this witty, well-written field guide, Richard Conniff studies the rich as a biologist studies the mighty mountain gorilla. In the process, he brings the rich down to earth as not only merely human, but distinctly animal."
—Frans de Waal, author of *The Ape and the Sushi Master*

"[An] unusual and delightful exploration of the richest members of the human species."
—*Publishers Weekly*

"The author surpasses Veblen in his analysis of the habits of the seriously affluent in matters of food, hobbies and sex, travel, nesting and sex, high jinks, foibles, inbreeding and sex. . . . A clever, invaluable zoomorphic study with a wealth of information on what makes the rich tick."
—*Kirkus Reviews*

"Mr. Conniff proves that his naturalist's approach to understanding the rich . . . really does generate insights and even understanding."

—*Dallas Morning News*

"Conniff [is] Truman Capote- and Dominick Dunne-like . . . a valuable field guide . . . endlessly fascinating." —*San Diego Union-Tribune*

"Very funny. . . . Recommended for libraries of all types . . . a fast-moving, instructive read." —*Library Journal*

"Of major significance is Richard Conniff's *The Natural History of the Rich*. Conniff conceived the brilliant anthropological notion to study . . . Wealthy People in natural habitats from Monaco and Paris to Aspen and Hollywood. What's best about this is that, not only is it a great idea, Conniff manages to pull off the ideal mixture of detached amusement and actual scientific fieldwork—so that his observations and conclusions are at once witty and sufficiently legitimate to avoid cheap satire."

—*The Day*

THE
NATURAL HISTORY
OF
THE RICH

A Field Guide

Richard Conniff

W. W. Norton & Company
new york • london

Because this page can not legibly include all permissions, pages 327–28 constitute
an extension of the copyright page.

For information about permission to reproduce selections from this book, write to
Permissions, W. W. Norton & Company, Inc., 500 Fifth Avenue,
New York, NY 10110

Manufacturing by The Courier Companies, Inc.
Book design by Brooke Koven
Production manager: Amanda Morrison

Library of Congress Cataloging-in-Publication Data

Conniff, Richard.
    The natural history of the rich : a field guide / by Richard Conniff.—1st ed.
        p. cm.
Includes bibliographical references.
    ISBN 0-393-01965-9
    1. Wealth—History. 2. Rich people—Conduct of life—History. I. Title.
    HC79.W4 C657 2002
    305.5'234—dc21   2002006899
    ISBN 0-393-32488-5 pbk.

W. W. Norton & Company, Inc., 500 Fifth Avenue, New York, N.Y. 10110
www.wwnorton.com

W. W. Norton & Company Ltd., Castle House, 75/76 Wells Street,
London W1T 3QT

1 2 3 4 5 6 7 8 9 0

Contents

Contents

Acknowledgments

For their enormous help with research, I would like to thank April Reese, Jane Maclellan, and Mark Urban at the Yale School of Forestry; Jenn Dinaburg at Connecticut College; and the staff of the Phoebe Griffin Noyes Library. Thank you to Roberta Frank at Yale University and my father James C. G. Conniff for help with fine points of language. To my wife, Karen, and our children, Jamie, Ben, and Clare, thanks for tolerance of chaos and help sorting it out. For their editorial criticisms, thank you to Fred Strebeigh at Yale University, Geoffrey Ward, Jim Doherty, and Bob Poole. For generously hosting me in the course of my research, thank you to Bob Chester and Nancy Larson, Amotz and Avishag Zahavi, Janice Braeder, the Hotel Sofitel Le Faubourg in Paris, the Hilton Hotel in Frankfurt, the Excelsior Hotel Ernst in Cologne; thank you also to Virgin Atlantic for help with travel. For their ideas and advice, thank you to Gilbert Tostevin at the University of Minnesota,

Sue Boinski at the University of Florida, Melissa Gerald at the University of Puerto Rico, Charlotte Beyer of the Private Investor Institute, Peter White of Citibank, Jay Hughes, and Alen MacWeeney. For assignments and encouragement along the way, thanks to Jane Berentson and Nelson Aldrich at *Worth*; Don Moser, Carey Winfrey, and Beth Py-Lieberman at *Smithsonian*; Steve Petranek and David Grogan at *Discover*; Jennifer Reek at *National Geographic*; Jason Williams and Bill Morgan at JWM Productions; and Jana Bennett at TLC and the BBC. Thank you to my agents John Thornton and Joe Spieler, whose antipathy toward Ralph Lauren caused him to suggest that I should make my natural history of the rich into a book, and to my editors Angela von der Lippe at W. W. Norton and Ravi Mirchandani at Heinemann. Thanks also to Stefanie Diaz for catching some of my mistakes.

By way of full disclosure, I must add that I have at various times done contract work, at several steps remove, developing ideas originated by CNN founder Ted Turner, Hollywood producer Peter Guber, and a computer software billionaire whose nondisclosure agreement says I will be turned to pixie dust if I reveal his name. My wife, Karen, and I were also once employed, near the bottom of the feeding chain, at a magazine owned by Kenneth Thomson. He killed the magazine in an act of mercy and put us out of work immediately after our marriage, encouraging me to begin my career as a writer. When I later visited him in Toronto, I found him to be a charmingly modest man. He even phoned ahead to his museum to make sure I wouldn't have to pay the $2 (Canadian) admission. I am grateful to them all, and if, in what follows, I occasionally bite the hand that feeds me, I trust they will regard it as a flea bite, and not the black death.

Introduction

Naturally Rich?

A telephone is to Armand what a fireplug is to a dog. He can't pass one without using it. The only difference is, he raises the receiver instead of his leg.

—Mrs. Armand Hammer

Let's begin with an embarrassing admission. Unlike certain pioneering works in the field of evolutionary psychology, this book did not have its origins on the rock-solid foundation of an attitudinal survey. Nor did it, like many great studies in animal behavior, emerge as a result of ten thousand hours of careful fieldwork watching spiders spin webs. It started with a tip from a stockbroker. I was visiting Monaco on an unlikely assignment for *National Geographic*, and it felt as if I had entered another universe, where even the most casual conversation was liable to veer off at any moment into the surreal. One day, for instance, I was having a friendly drink with two young women seeking starter husbands when one of them asked the other, "Does he still have the Jaguar with the matching dog?"

"It was a Morgan," her friend replied. "Cream colored."

I asked one of them to teach me some French, and the phrase that came dancing gladly to her lips was *"Il a du fric,"* or "He's loaded." I, on the other hand, could not even open a savings account in Monaco, where the bankers gently informed me that they had a $100,000 minimum.

"National Geographic?" a British stockbroker asked me one morning. "Why aren't you in the mountains of Papua New Guinea?"

I suggested that every nation has its own anthropology, and that the native customs of Monaco were at least as exotic as those of any other hill tribe.

The stockbroker latched onto this idea instantly. "You know, I go to this nightclub called Jimmy'z, and it's like a ritual. Every night they play the same songs, and the same girls jump up and . . ."

I ran into the stockbroker again a few nights later at a piano bar, and he immediately began cataloguing the Monégasque anthropology all around him: "Here you have the tribal mating ritual. Everyone out on display. Lots of plumage. Lots of cross-fertilization."

"You sound like David Attenborough," I said.

He was holding his drink up in two hands, and the fingers flapped open to indicate the brightly clad woman on the next barstool. "And here," he said, in a hushed natural history presenter's voice, "is the black-and-white striped, red-collared ladybird."

Just at that moment, in the smoky mirror opposite, I noticed a sleek man with an abundant mustache preened out past the corners of his mouth. He was in profile, like a side character in a Toulouse-Lautrec painting. He held up his gold lighter and, loving every gesture, stroked the flame around the tip of a fat cigar. Then, with this priapic accessory firmly in hand, he passed about the bar happily greeting his many acquaintances. In this glittering *über*world of trophy wives and international arms dealers, everyone was rich and beautiful.

The stockbroker sipped his drink and had second thoughts. "We're all the same beast, with or without the Cartier," he said. "You can mess around with the signs and symbols, but it all comes down to the same thing as the monkey's red arse, which is, 'Pay attention to me.' "

This idea took hold in my imagination, possibly because my life had been meandering in roughly this direction for some time. I have spent much of the past twenty years writing about the natural world, for magazines like *National Geographic* and *Smithsonian*. At the same time, I have regularly contributed articles on distinctly unnatural-seeming subjects to *Architectural Digest*. So I have alternated between writing about the rich and about the bestial, often in close proximity. My assignments have taken me from champagne with Richard Branson at a private club in Kensington to a casual swim with piranhas in the Peruvian Amazon, and from an interview at Blenheim Palace with John George Vanderbilt Henry Spencer-Churchill, eleventh Duke of Marlborough, straight onto a plane to an audience in Botswana's Okavango Delta with an aristocratic baboon named Power. It was a toss-up which of these worlds was more perilous and, traveling between the two, it was impossible to avoid seeing certain similarities. When I got to Botswana, for instance, a biologist I was visiting confided, "The rules with baboons are the same as in a Jane Austen novel: Maintain close ties with your relatives and try to get in with high-ranking animals."

As a natural history writer, I've always assumed that all individual animals, from the Australian bulldog ant on up to Rupert Murdoch, conform, more or less, to the rules of their species. They fit into basic patterns of physiology, territoriality, social hierarchy, reproductive behavior, parental care, and so on, and the ones that don't fit generally get eaten.

So I began to wonder if it wasn't possible to think of the rich in a new light—as animals, that is. It occurred to me that there might even be a natural history of the rich, something along the lines of "The rich are different from you and me. They do more scent-marking." The German poet Heinrich Heine, for example, once described a visit to the Paris offices of Baron James de Rothschild. When a "gold-laced lackey" exited the Baron's private rooms bearing the great man's chamber pot, a stock market speculator waiting in the anteroom actually stood and "reverently lifted his hat." It reminded me, naturally, of the obsequious behavior of subordinate house mice in the presence of a dominant male's urine.

Likewise, when Revlon billionaire Ronald O. Perelman strongly discouraged his third wife, Patricia Duff, from attending parties solo during the 1996 Democratic Convention, it suggested the mate-guarding behavior of certain wasps. On finding a desirable female, a male wasp will often grab her and hustle her into hiding to prevent other males from mating with her.

This sort of analogy may seem outrageous and demeaning, so let me acknowledge up front that one can take this only so far. It would be wrong, for instance, to liken Perelman to the fly species *Johannseniella nitida*, in which the male's genitalia break off after copulation and become a sort of chastity belt on the female. On the other hand, the rich themselves liken their own behavior to animals with a frequency and persistence indicative of some deep yearning to know their place in the natural world. John D. Rockefeller, preparing to launch one of the most predatory raids in the history of American business under the innocuous guise of the South Improvement Company, confided to his wife, "I feel like a caged lion and would roar if it would do any good." Nor are these analogies always along such predictable lines. Robert Dedman, Sr., a member of the *Forbes* 400, once described the joy of acquisition in these terms: "I feel like a mosquito flying over a nudist colony wall. It all looks so good I don't know where to start."

But is it plausible to talk about a natural history of the rich? Can we discuss the natural behaviors of a socioeconomic class when the boundaries between classes are so flexible, and when a Bill Gates can go from upper-middle-class dweeb to world's richest no-neck oligarch in a decade or two? Or, for that matter, when an Internet magnate like Masayoshi Son can lose $63 billion in a single year and still have more money than God?

On the face of it, the idea was ridiculous. The rich *aren't* different from you and me genetically (setting aside, for the moment, certain cases of assiduous inbreeding). We are all, as my Monaco stockbroker put it, the same beast. We all developed our essential biological and mental mechanisms for dealing with one another and the world back when the species was still dragging its knuckles in the dirt—and those ancient adaptations still shape the way all of us behave to this day. We evolved to gorge on fats and sugars, for instance, as hunters out on the savanna, where these essential resources were scarce—and we're still gorging on them as office workers, even though they're now abundant enough to give us heart-woe and tooth-rot.

But the rich get more candy than the rest of us. Or rather, they get more of whatever it is they happen to want, at any given moment. Whenever any animal gets more of a resource, it has a way of changing the animal's behavior. Give a howler monkey more food and higher status and he will tend to have sex more often. Give an opossum more well-stocked garbage cans and she will soon start to skew her birth ratio, producing more sons and fewer daughters. Individuals with roughly the same genetic makeup behave differently, given different resources. "Everyone knows," the primatologist Frans de Waal has written, "that a sunflower seed planted in a sunny spot in the garden gives a completely different plant from a seed planted in the shade."

Might this apply to humans, too? At the very least, our shared behavioral and biological tendencies matter considerably more when they express themselves in the lives of the rich. No one much cares if the average middle manager lowers his brow and storms around trying to assert his status. But when Bill Gates displays the same sort of dominance behaviors, companies die, people get rich or go broke, and the U.S. government, among others, begins to pay close attention.

Looking at rich people from an animal-behavior perspective might be revealing and useful even for the rich themselves. Gates, for instance, might never have gotten into such trouble with the U.S. government had he understood the subtleties of dominance behavior as practiced in the animal world. It's Chapter One in *Leadership Tools of Tamba the Ape*: Chimps will sometimes assert their status by ripping out a rival's eye. But they also practice dominance by the gentler means of social manipulation, alliance building, and even by being nice.

Zoomorphism (looking at humans as if they were animals) and anthropomorphism (looking at animals as if they were humans) are both ancient pastimes. But a number of developments over the past few decades have shown that people and animals have a lot more in common than we thought: The most important is that biologists like Jane Goodall, Amotz Zahavi, and Gordon Orians have begun to present the results of the first detailed, long-term field studies of animal populations in the wild. It might be difficult to see the connection between a rich woman swanning around in her Manolo Blahniks and some underpaid clipboard-wielding biologist slogging through the bush in battered Tevas. But this is the first time anyone has ever followed the life histories of known individuals in the animal world over years and even decades, recording, among other things, who their grandparents were, how they gain or lose power, how their rank affects mating behavior, and how they pass on their status to their offspring.

At the same time, scientists have begun to acknowledge that animals think and feel, a notion that was biological heresy as recently as the 1970s. They are sorting out just how animal thinking resembles ours and how it differs. Molecular biologists have meanwhile demonstrated how closely our genetic heritage resembles that of other species. It's now common knowledge that the human and chimpanzee genomes are 98.4 percent identical. But even when scientists look at a species as different from us as the tiny, wormlike nematode *Caenorhabditis elegans*, they find that 74 percent of the gene sets—the basic working units of the genome—have close counterparts in, say, Martha Stewart. It has become increasingly evident that humans and other animals are part of a biological continuum. For all her apparent profligacy in creating new species, Nature has turned out to be a bit of a cheapskate, using the same tricks over and over, from one species to the next. This parsimonious tendency at times defies belief, as when the pheromone of the cabbage looper moth also turns up as the chemical signal for sexual readiness in the Asian elephant.

These biological findings have coalesced in the new field of evolutionary psychology, which examines how our early history on the African savannas continues to shape our behavior on Sutton Place and the Avenue Foch. Evolutionary psychologists of course look at behavioral and biological forces that shape the way we live in Perth Amboy, too. But researchers studying any species pay special attention to the dominant individuals, and in the case of the human animal, that generally means rich people. Studies suggest that some distinctly unsavory survival mechanisms, and a few rather more savory ones, express themselves with special intensity among wealthy people, if only because there's so much more at stake. Hence the natural history of the rich.

Evolutionary psychology has thus far provoked two obvious criticisms. The first is that it merely reinforces traditional sex roles: Males get money and multiple sex partners. Females get the double standard. The conventional argument is that the Armand Hammers and Nelson Rockefellers of the world tend to be unfaithful because males gain a huge biological advantage, and incur so little cost, in spreading their seed as widely as possible.

But an evolutionary predisposition to infidelity may operate just as powerfully, though perhaps more selectively, in females. Biologists now recognize that the female bluebird, long thought to be an exemplar of lifelong monogamy, often indulges in what they delicately term "EPCs," or "extra-pair copulations." While her mate back at the nest may be a good provider, he is, all too often, dull. So she turns to her lover for gifts, the thrill of amorous attention, and possibly also for better genes. In one study, more than 15 percent of a bluebird's offspring turned out not to have been fathered by her nestmate. What could this possibly have to do with the lives of the rich? In much the same manner as nesting bluebirds, Jennie Jerome and Randolph Churchill shared a marriage of moderate wealth and aristocratic stature, of which Winston Churchill was the first-born son. But Lord Randolph suffered from untreatable syphilis, and Jennie went on to take more than two hundred lovers over the course of her life. Lieutenant Colonel John Strange Jocelyn appears to have been one of them, in the summer of 1879. The following February Lady Churchill gave birth to Winston's brother, who was named John Strange Spencer Churchill.

A second possible criticism is that a natural history of the rich may sound like a return to social Darwinism. It has become part of our intellectual mythology that the robber barons of the late nineteenth century liked to use Darwinian survival of the fittest as a shibboleth for praising themselves and bashing the poor. Rich people then supposedly took comfort in the notion that their wealth represented, in John D. Rockefeller, Jr.'s, phrase, "the working-out of a law of nature and a law of God." But the phrase "survival of the fittest" entered the language well before the 1859 publication of Charles Darwin's *The Origin of Species*; it was coined by the economist Herbert Spencer. Moreover, rich men then were far more likely to attribute their success to God than to nature. John D. Rockefeller, Sr., specifically extolled the ostensibly cooperative and Christian character of his Standard Oil trust as an antidote to Darwinism.

In any case, I hope to use a more elegant modern form of Darwinism neither for praising nor for bashing the rich, but for understanding them and explaining how social patterns established when their forebears were grunting around ancient campfires continue to influence their behavior in the modern day. In this context, Darwinism gives us a sort of field guide to the behaviors and strategies of the rich: What evolutionary mechanisms enable them to acquire wealth? How do they use primate dominance behavior to preserve their wealth and translate it into status? Why is the annual Rockefeller gathering at their Pocantico Hills estate not just a family reunion but a Darwinian instrument for maintaining a dynasty? What nuances of display behavior compel the rich to travel to the same destinations—Aspen and St. Moritz, Nantucket and Majorca—and meet the same people, over and over? I do not intend to make moral judgments about the rich, any more than I would put a moral value on, say, the tendency of female prairie dogs in the last harsh days of winter to kill and eat their sisters' offspring or, for that matter, on their tendency in the first glorious flush of spring to nurse those offspring as if they were their own. A naturalist learns to judge different strategies only according to their survival value.

My intent in this book is to use the tool of evolutionary psychology with gleeful caution. My brief as a journalist is to doubt, to provoke, to inform, and to entertain. I will at times go well beyond anything scientists would assert. If I liken Donald Trump to a hangingfly, for instance, readers may surmise that I am edging off the path of rigorous scientific inquiry into mere irony and speculation. At the same time, I will of course note when the evolutionary psychologists themselves skip off into the absurd. One writer has suggested, for example, that passionate kissing was invented by rich people in the Middle Ages, as part of a system of courtly love designed to delay marriage and reproduction among the landed gentry. But surely Salome liked a good snog? Or Cleopatra? Or even Lucy?

My point of view will be at heart sympathetic, for at least three good reasons. First, we are all probably descended from the rich. This may seem unlikely at first. The most notorious assertion of Darwinism, after all, is that we are descended from apes. But recent studies suggest that we are almost certainly also descended from kings. In Darwinian terms, we are descended from dominant animals who have parlayed their social status into reproductive opportunities. Not long ago, an enterprising architectural preservationist wrote to fifty of the richest people in Britain seeking their support as "descendants of King Edward III." It was a safe genealogical gamble because Edward III, who reigned in the fourteenth century, had seventeen children, giving him roughly two million descendants in modern Britain. Descent from King Henry I, who reigned in the twelfth century and fathered perhaps fifty illegitimate children, would have been an even safer bet.

Second, we all hope to be rich ourselves. We are descended almost by definition from people who liked food and sex. From them we have inherited deeply embedded biological drives for status, for waterfront real estate, for landscapes of the English country house variety (derived ultimately from the African savanna), and for a variety of other attractive features often associated with the rich and famous. Moreover, all our disclaimers to the contrary, we long to be like them. We pay attention to the rich as slavishly as a troop of gorillas following the lead of its dominant silverback. We mimic them as aptly as a viceroy butterfly mimics the coloration of a monarch. As in any dominance hierarchy, we also fear the rich. They can use their power to hurt us in ways we hardly recognize.

Third, I intend to be sympathetic to the rich because, like bull elephants in musth and hungry grizzly bears in hyperphagia, they have their own peculiar plight. Inbreeding may be an advantage because it allows them to concentrate family wealth, but it also increases the threat of imbecilism and disease. (For example, the porphyria that afflicted George III and probably also Kaiser Wilhelm II.) Marriage to an heiress may be a useful tool for shoring up the dynasty, but it also entails a heightened risk of infertility.

Moreover, life at the top can be in some ways exceedingly unpleasant. I recently heard a trust fund beneficiary rattle off a list of woes associated with inherited wealth, including social isolation, resentment from peers, rich-bashing from society, betrayal or exploitation by friends, unrealistic expectations from family and society, unequal financial status in marriage, and an absence of all the usual factors (like worrying about the rent) that cause the rest of us to drag ourselves out of bed most mornings in search of bread and a modicum of self-worth. He might also have added that the rich get no sympathy. When you are worth millions of dollars, there is a presumption that you should shut up and bear it. The rich do not necessarily even find much consolation in one another, because the sprawling, walled-off character of their homes isolates them from members of their own families. About her upbringing on the family estate ten miles outside Louisville, Kentucky, Sallie Bingham (of the newspaper Binghams) wrote, "Alone on the Place, we were a tribe, as isolated and peculiar in our habits as any group of Stone Age creatures in the forests of New Guinea."

Finally, life at the top is intensely competitive. Wealth often drives the rich to extremes of display and defensive behaviors, much as a bull elephant seal must spend a thousand pounds of blubber every mating season to reinforce his status. Appearing on the *Forbes* 400 list of the richest Americans did not give Ted Turner a sense of having arrived: "The first time I looked at it and saw my name, I thought 'Hmmm, I can do a little better.' " Yet striving to do a little better can lead all too easily into *folies de grandeur*, as when a wealthy New York couple threw a party in rented rooms at Blenheim Palace, the ancestral seat of the Churchill family, and sent out engraved invitations reading, "Mr. and Mrs. John Gutfreund, At Home, Blenheim." The world is always avidly watching the rich, careful to note if the sound of their chest-thumping is persuasively resonant, or merely hollow.

With this book, I am proposing that we listen a little differently, traveling into the world of the rich as if we were anthropologists making the first visit among the tree-dwelling Kombai tribe of Irian Jaya or a primatologist among the squirrel monkeys, alert to what is lovely, poignant, and also ludicrous about their lives.

It was in this spirit, on my Monaco stockbroker's advice, that I headed out one Saturday night to the nightclub known as Jimmy'z. At 1:00 A.M., there were five Bentleys, three Rolls-Royces, six Ferraris, and too many Mercedes to count lined up in a neat hierarchy by the entrance. It was an ordinary Saturday night. My guide palavered with the local chieftain. (His name was Franco. There was no Jimmy; the name of the place is just a hip-sounding pseudo-Americanism.) We were soon seated at one of the tables reserved for Monaco regulars, by a low wall overlooking a pond.

The windows had been rolled away for the night, so the room was open to the Mediterranean breezes. A bridge led across the pond to a Japanese garden on the other side. When the weather is right, people get up and dance on the wall. Sometimes they also fall into the pond. A glass of champagne cost $40. But champagne is practically Monaco's national beverage, and a glass of mineral water was only $2 less, so staying sober did not pay.

I sipped my drink, and faces from the village began to materialize out of the darkness. My British stockbroker went sweating past, intent as a great white hunter on some unseen quarry. I discerned the hierarchy of tables radiating out from the ones reserved for Prince Albert and his friends. I recognized a real estate tycoon and a Saudi oil magnate. A woman in a sheer bodysuit jumped up onstage, and though her face was new to me, her learned demonstration on primate mating behavior was deeply familiar.

A little later in the night, I found myself out on the vaguely psychedelic dance floor. All around me, the cream of European civilization, representing an inordinate share of the human wealth and beauty of the planet, jostled and glistened to a song called "Don't Want No Ugly Mothersucker." (Flip side of the little-known crowd-pleaser, "Don't Want No Short-Dick Men," the two songs together constituting a brief but cogent treatise on evolution by sexual selection.)

I spotted a demure Asian woman in a traditional white silk dress, passing by with delicate steps, hemmed in by four bodyguards. She was giving a small party on the other side of the pond and sending over drinks to the locals, who buzzed furiously about whether she was Chinese, Japanese, or Thai, and whether she was, as rumored, a princess. No one knew, and it was necessary ultimately to fit the whole thing into the tribal hierarchy: "It doesn't matter who she is. She's nobody in Monaco, because she's sitting on the other side of the pond. Anybody who's anybody sits on this side at Jimmy'z."

Soon after, I caught my plane home, happy to be nobody in Monaco, and yet also tantalized. In this world of diminishing wonders, I had found a new quarry, and they were possibly the most dangerous and elusive animals on Earth. All the familiar tools of biological research—the radio collars and tranquilizer darts—would be of no help to me in this new hunt. I was not going to be sampling anybody's spoor. The habitats I needed to explore would be strange and in some cases forbidden.

Back home, I looked longingly at the web vest and the battered Gore-Tex boots in my closet. Then I picked up the phone and, drawing strength from the spirit of intrepid explorers past, I booked a flight to Los Angeles and a car, a bright red Ferrari F355 Spider convertible worth $150,000. It was a rental, to be sure (and with a $10,000 deductible, at that). But I figured that for the adventures ahead I would be needing very different kinds of camouflage.

The Natural History
of the Rich

1

Scratching with
the Big Dogs

How Rich Is Rich?

There's a certain milieu in Aspen. To the extent that you're collecting important art, you're listening to avant garde music, and you participate as a peer in Aspen Institute intellectual discussions—that, and you have $100 million—then you're considered a big dog.

—HARLEY BALDWIN, *art dealer*

IF MEN COME FROM MARS AND WOMEN FROM VENUS, WHERE ON earth do rich people come from? Are they, as ordinary people often suspect, an alien life form? Is their blood the color of money? Do they have special antennae, as their press people like to suggest, for picking up distant intimations of profit and loss? Can they see around corners? Is life on Canis Major, the big dog star, really light years apart from the bow-wow world of ordinary runts like you and me? The truth is that rich people are not even a different species from us. They are more like a different subspecies.

The rich themselves often say that they just want to be normal people, leading normal lives. "I just want to be middle class," was a familiar refrain among dazzled Internet millionaires in the late 1990s. Then, to their horror, they got what they wished. This ambivalence about wealth is perhaps sincere, but it's also a little disingenuous. Jeff Bezos of Amazon.com made himself a folk hero of the era as a billionaire who drove a beat-up Honda and celebrated frugality. "I don't think wealth actually changes people," he declared. But at the time he was moving out of his 900-square-foot rental in downtown Seattle to a $10 million waterfront house in the leafy suburb of Medina, where his new neighbors included Microsoft billionaires Bill Gates, Jon Shirley, and Nathan Myhrvold. Then, 7,000 square feet perhaps seeming relatively frugal in this context, he decided to expand the place. Wealth is like that.

Whether they want it or not, the dynamic of being rich invariably sets people apart. It isolates them from the general population, the first step in any evolutionary process, and it inexorably causes them to become different. They enter into a community with its own behaviors, its own codes, its own language, its own habitats. ("I'm the most normal, normal person," one extremely wealthy woman told me. "I'm not like most rich people. I work really hard. Most rich people I know don't do anything but eat, drink, sleep, pardon the term, fuck, and have a good time.") Their children or grandchildren come to mate mainly with one another, like Whitneys with Vanderbilts and Firestones with Fords. (If you are planning to get a wedding present for little Jennifer Gates Bezos, start saving now.) Thus, from the primordial muck, something new and wondrous emerges: A cultural subspecies, *Homo sapiens pecuniosus*.

How to identify the breed? Is there a holotype, a specimen pinned down in a museum somewhere against which one can size up all new-comers and say this is or isn't a rich person? Is it really possible to charac-terize a group that includes both a comparatively dainty figure like French businessman Bernard Arnault, purveyor of Louis Vuitton and other opulent brands, and a bruiser like basketball star Shaquille O'Neal, who weighs 330 pounds and has the word "TWISM" ("The World Is Mine") tattooed on his left bicep? It is. The way to start is by defining just what we mean by "rich."

A Numbers Game

One afternoon in Aspen, I had coffee with a local craftsman. He was the second person that day to let me know early in the conversation that he didn't need to work for a living. He'd married into a prominent family, and when the name failed to produce a satisfactory response, he said, "They owned General Dynamics," a manufacturer of some of the deadliest weapon systems on Earth. "They owned the Empire State building," he said. "Do you have the *Forbes* 400 list?" he asked. It turned out that they are currently worth about $3 billion.

He was a solid, muscular guy, with an upright, almost balletic posture, and a manner of quiet arrogance, in the first person plural. "We really, *really* go out of our way not to dress particularly well, not to drive really fancy cars—everybody in the world can get a Range Rover—and not to let people know what our philanthropic endeavors are. Sometimes you don't want the advertisement."

He was scathing about wannabes. Maybe it was because he was himself a relative newcomer in this world. "You can't pretend to have the speed of a cheetah, when you're really a mule," he said. A new country club in town especially irked him. It created "a different level of Aspen citizen, those who belonged, and those who didn't. It was really terribly exclusive in a way a lot of us resented." He'd signed on as a charter member, just to get in a quick round of golf. But the other members turned out to be, on average, sixty-four years old. Mules, not cheetahs. They needed five-and-a-half hours to complete a round. So having bought his membership at $60,000, he sold out at $175,000 and could savor his righteousness.

He asked, as everyone did sooner or later, how I was planning to define wealth and I said that I was probably going to make my starting point between $5 and $10 million in investable assets. "I don't consider that a lot of money at all," he said. This was an entirely reasonable comment, improbable as it may seem. For $5 or $10 million, you could just about buy a suitable home in Aspen, where the average house was then selling for $3.4 million. (It would be a second home, of course, so you would also need funds to feather it, to fly back and forth, and to entertain your fine new neighbors.) In any case, he added, money doesn't matter: "Money by itself doesn't interest me all that much." He defined wealth essentially as contentment with your lot: "My opinion of wealth is to be able to own whatever it is you have, at whatever level. If you have $50 million and you're racing around bloody well killing yourself and basically a slave to that which you are striving for, I would not count that as wealth in any shape or form." I demurred. The guy with $50 million may be a slave to it, but people still leap at his bidding.

"So this book is just going to be a numbers game, then?" he said. "It's about what's in my billfold?" He took out his billfold and showed it to me, to demonstrate how crass I was being. There was a dollar on top, evidence of how much he cared about understatement. Then he peeled it back to show me the $100 bills underneath.

I picked up the check.

Money Doesn't Interest Me

It was a reminder of just how strange and complex an animal I had set out to study. Not just old money or new, but old money with nouveaux riches husbands, and trophy wives who turn around and trade up from Big Daddy to Bigger. Working rich, of course, and idle rich. First-generation tyrants and fourth-generation wastrels. Rich people who read Epictetus and honestly wonder, What can I do to make this a better world? And rich people who mainly wonder, Who can I crush today? Seattle rich who keep quiet, and Los Angeles rich who get out of bed to a crescendo of timpani. New money on polo ponies, and old money on roller blades.

What do they all have in common? Almost all in one form or another expressed the idea that money by itself didn't interest them that much. In the beginning, this sounded like the fourth biggest lie, along with "the check is in the mail," and so on. If so, it was a lie with a great tradition. In the library at The Breakers, their seventy-room cottage in Newport, Rhode Island, for instance, Cornelius and Alice Gwynne Vanderbilt had a white marble mantle bearing the venerable French inscription, "Little do I care for riches, and do not miss them, since only cleverness prevails in the end." Biographer Barbara Goldsmith writes that the Vanderbilts saw no irony in purchasing this mantle, which had been pried off the fireplace of a 400-year-old château in Burgundy. Presumably the builder of the château also saw no irony in putting the mantle there in the first place. Rich people have always believed it is their cleverness, their wit, their taste, their athletic ability—anything but their money—that makes them special.

And yet they often acted as if money was the only thing that interested them. They practiced the dull art of price-tag parlor talk: "The trouble with Arnie is that he'll only spend $150,000 for a pilot, when he could get a damned good one for $250,000." They applied price tags with wild, domineering abandon even to the most delicate questions of marriage and family life. A photographer friend who was making portraits of two gorgeous younger wives not long ago overheard one of them discussing a sex act proposed by her aging husband. For better and for worse, the details of this sex act are unknown, except that she refused to participate. So he offered her $100,000 and then $200,000. "I'm not doing it," she said, to which he replied, "$350,000, and that's my final offer." She thought about this for a moment, perhaps contemplating what her mother once told her about the spirit of give-and-take in marriage. Or maybe she was just thinking about the price of a Russian sable fur coat. Then she said, "I'll do it for that."

Being a Big Dog

Yet the phrase "money doesn't interest me" got repeated so often, and at times with such sincere boredom at the idea of another million gained or lost, that I began to think it might not be such a big lie after all. Money, that is, real money, was of course essential. But at least after the first flush of pure cash bedazzlement, most people discovered that money by itself wasn't enough. What you could do with it, on the other hand, what you could become, was endlessly fascinating. "Make no mistake," a private family banker declared one day, on the topic of what rich people are seeking with their wealth, "what this is all about is love." No doubt he was right, up to a point. But my background as a naturalist inclined me to think it was more often about the things that drive top animals in the natural world—the quest for control, dominance, mating opportunities, and, above all, status.

Money merely gets you into the game, and the price of admission can vary wildly depending on the milieu. So it was much more complex than it might at first seem to ask, What does it mean to be rich? or How do you define wealth? In the Himalayan nation of Bhutan, I went trekking with a prince, a first cousin of the king, who had no money in the American, much less the Aspen, sense. Yet people covered their mouths when they spoke to him, lest they taint him with their human breath. He carried the talismanic aura of his cousin, and thus he was rich in at least one classic sense: The word "rich" derives from the same Indo-European root that produced the Celtic word *rix,* the Latin *rex,* and the Sanskrit *rajah,* meaning "king," and in many cultures the concept of being rich is about how closely one can approximate the aura of royalty. My Bhutanese prince may also have been rich relative to local economic standards. If, as a fellow trekker suggested, you begin to be rich when you earn twenty or thirty times the local per capita income, then in Bhutan, where the per capita income is $510, someone with an income of $15,000 could conceivably live like a king.

And in the United States? "A fortune of a million is only respectable poverty," a prominent figure in the New York 400 remarked in 1888. Yet the word "millionaire" has held onto its own talismanic cachet. *Who Wants to Be a Millionaire* became a television hit despite Regis Philbin. *The Millionaire Next Door* became a bestseller even though it featured timid souls who stayed at home on Saturday nights, drove Buicks, and held J.C. Penney credit cards. These millionaires actually had a median income of $131,000 and were the sort of folks who could be motivated by a $200 fee to sit down for "personal and focus group interviews." They were, in short, working stiffs, with a relatively modest net worth.

Junior Wealth

For the purposes of this book, immodest wealth is more the idea. A recent survey prepared for the financial services industry found that there are 590,000 "pentamillionaires" in the United States, and the authors predicted, with a gleeful rubbing of hands, that there would be 3.9 million of them by 2004. This economic cohort consists entirely of people with a net worth, not counting the primary home, of $5 million and up. Other reports bandy about terms like "centimillionaire" and the rather dazed and dreamy "gazillionaire" or even "kabillionaire."

But the word "millionaire" still glimmers in the public imagination, even if that means we must drastically revamp it to fit modern times. In a recent survey by *The Wall Street Journal*, most people defined wealth not by assets but by an *income* of a million dollars a year. This is heady territory. In 1999, the Internal Revenue Service received only about 205,000 tax returns with an adjusted gross income of $1 million or more (most of them joint returns, at that). It begs the question: Could you be rich on somewhat less? That is, could you arrive at the nebulous sweet spot where you are doing what you like—and what your neighbors wish they could be doing—without having to worry much about how to pay for it? (You might think that being rich means no longer caring what the neighbors think, but the art of wealth is at least partly about the ability to inspire and manage envy. If what you like to do is sit in the attic and clip dolls out of yesterday's newspaper, you may have ownership, you may have contentment, but you are not rich. Do it with $100 bills, on the other hand. . . .) So what is the magic number that makes you rich? Based on no surveys or economic parameters whatsoever, other than a brief look at this month's bills, my own impression is that an income of somewhat less than $1 million a year would do quite nicely. A salary of $500,000 might not make you rich, because of the obligation to punch a clock and kiss the hierarchical hinder parts, but an income at that level from one's own portfolio begins to sound pleasantly independent. The threshold for wealth thus seems to me (and to most private bankers and big-time fund-raisers) to lie somewhere in the unpoetic but highly palatable pentamillionaire zone. The exact number is subjective. It depends on what you like to do and where you like to do it. In Aspen or Palm Beach, an investment portfolio worth $5 or $10 million will certainly not cause people to drag themselves to their feet for a round of curious sniffing. In New York, bankers refer to a fortune of $10 million as "junior wealth." You need at least that much, and probably far more, to keep the apartment on Park Avenue, the place in the Hamptons, and the kids at Brearley or Spence. But in most places in the United States, Europe, and Japan (and therefore also in less developed nations), it would give you a fair start at being a big dog.

Wait! Sit! Before we head off in pursuit of the rich, we need to think about two related concepts critical to the definition of wealth and the character of our subspecies. The first concept, relative deprivation, has to do with the point at which the rich themselves feel they are becoming big dogs; the second, social isolation, is about how they choose the habitats where they typically go to say, Woof, woof!

Relative Deprivation

It is part of the elusive nature of wealth that people with money seldom think of themselves as rich. Or at least they are careful not to say the words "I'm rich" out loud, possibly not even to their spouses. At a brasserie in Paris not long ago, I met a lovely, unpretentious woman who spends all her time in the company of the very rich. To get a sense of her perspective, I asked if she had grown up middle-class. She hesitated, genuinely uncertain. Then she described her childhood: A house in Paris, a driver, a chef, a maid, a house in the south of France for July, and a house in Normandy for August, with Monets and Sisleys on the walls.

"You had to think about that?" I asked.

"In France," she explained, "we have the idea of, not middle class, but *grande bourgeoisie*." She was, in any case, quite certain that she is not rich now.

Hardly anyone is. In a recent survey of people with a net worth between $1 and $4 million, for instance, only 9 percent would admit to being wealthy. (The word "rich" was apparently too raw for even the pollsters to utter.) The rest said they were comfortable, or possibly "very comfortable." About half the survey respondents defined wealth as $5 million or more.

When I talked to people in the course of my own research who were worth $5 million, the magic number was more likely to be $10 million. For people with $10 million, it was $25 million, and so on ever upward. For some people, no amount of money is ever enough. When Commodore Cornelius Vanderbilt was suffering from stomach pain shortly before his death, his doctor prescribed champagne. "I can't afford champagne," Vanderbilt replied. "I guess sody water will do." At the time, in 1876, he was worth $110 million, $5 million more than the Federal Reserve. Wealth is often like that, beckoning from some place just out of reach.

Nelson Peltz, for instance, made his fortune as the eager pawn in leveraged buyouts orchestrated by Michael Milken in the 1980s. He now lives on a 130-acre estate called High Winds in Bedford, New York, and likes to irritate his wealthy neighbors by hopping into his Sikorsky six-seater to run down to the corner store. (OK, this is a lie. He is so concerned about the peace and well-being of his neighbors that he never uses the helicopter for anything less important than avoiding the tedious five-minute drive to Westchester County Airport.) Peltz is now worth $970 million, which makes him a centimillionaire—a fine thing to be, but just painfully shy of the magical "b" word. "You see these guys worth $3 billion to $4 billion," he recently lamented, "and you think to yourself, 'What have I done wrong?' "

Psychologists call this "relative deprivation," the tendency to evaluate oneself not by objective criteria, but by comparison with a select group of peers. It is why even the superrich often do not think of themselves as rich: Because there is always somebody a little richer just up ahead, or somebody who threatens to become richer toiling close behind. Hence John D. Rockefeller's remark, on learning in 1913 that J. P. Morgan's estate was worth a mere $80 million: "And to think he wasn't even a rich man." Odd as it may seem, the rich gravitate toward one another in their exclusive enclaves partly to facilitate these comparisons. The pursuit of a forum for inflicting a sense of relative deprivation, and also for enjoying a sense of relative *comfort,* is one reason the rich are prone to social isolation.

Splendid Isolation

One day in Los Angeles I turned off Sunset Boulevard and drove up the winding, verdant roads into Bel Air to visit the home of the Hollywood producer Peter Guber. Guber, who made his biggest box office splash with the film *Batman,* lives in a twelve-acre hilltop compound at the end of a cul de sac, behind a high masonry wall with a massive pegged-oak gate. It could be the entrance to the Batcave. When I identified myself via the security speaker, the gate swung wordlessly open. A driveway circled up to the main house, where an assistant walked me through another pegged gate into the courtyard, through the polished marble hallway, past a Dubuffet, and finally through yet another pair of gatelike doors into the pool room. There I waited, beneath a Hurrell portrait of Greta Garbo, who was staring down rather hard. On the coffee table was a book with a Charles Bell painting on the cover, depicting the innards of a pinball machine, much like the large Charles Bell painting on the wall. Another book, on Louis Comfort Tiffany, served notice that the Tiffany table lamp with dragonflies on the shade had not come from Pottery Barn.

Guber was a tall, lean, animated character, in blue jeans and a polo shirt, with a small feather that had somehow gotten caught in his hair. He was obsessed at that moment with the question of "shoot-outs" in the entertainment business, the topic of a book he was working on. He was also nursing a wounded reputation, because he'd been a big dog during Sony's disastrous entry into Hollywood and was said by one book to have helped perpetrate "the most public screwing in the history of the business." It made the idea of living behind stone walls seem prudent. "If people know you're a killer, they want to kill you," he said, at one point. And again, "Sometimes there's a shoot-out, and you don't realize it's happened. You walk home or you go to a restaurant, and you fall over in your soup."

Guber was not only a combative character but also highly territorial. He had a 200-acre oceanfront place in Hawaii and a 1,000-acre ranch in the Woody Creek section of Aspen, with signs out front threatening to close the Nordic ski trail if people used his driveway to get to it. In Los Angeles, his house looked down on the celebrated Hotel Bel-Air, "where every guest becomes part of the legend." Guber had negotiated to build a long granite stairway down to the hotel with a private entrance, so he would not have to linger among the mere legends at the front door. The deal also included 24-hour room service at home, in case the urge for a salmon tartare with caviar and marinated cucumbers should ever strike at some dark hour. Yet, at one point in our conversation, Guber remarked, "The trouble with being rich is, you only meet other rich people."

On some level, he probably meant it. But of course the rich routinely structure their lives for the specific purpose of meeting only other rich people. It's why the same people turn up on the same big weekends in Aspen, making the airport so crowded that the Learjets must huddle like fledglings under the wings of the Gulfstream Vs. It's why you can find billionaires lined up at 4:30 A.M. during Christmas week to get a chaise by the pool at the Four Seasons Maui (the Four Seasons has somehow devised a system whereby one cannot bribe or send a substitute to hold one's seat; you have to show up yourself before dawn). It's why the people who summer together on Fishers Island also winter together in Boca Grande, and why the people who lunch on Tuesday at Le Cinq in Paris may recognize their fellow guests on Friday afternoon at the Hôtel du Cap on the Mediterranean. They arrange their lives so that, wherever they go, they see the same few hundred people. The people who matter. It can seem as if no one else in the world even exists.

A Financial Freak

They do it at least partly because a rich person is a kind of freak in the world at large. "The mailman, the lady at the dry cleaners—they look at me with a price tag," says Leslie Wexner, the retailing billionaire. "I see it in their eyes. I'm a financial freak." People gawk in public places, old friends appear out of nowhere, and everybody wants something, if only the magically enhanced pleasure of one's company. Gertrude Vanderbilt Whitney, for instance, was an immensely rich woman whose deepest wish was to be known not for her money but for her talent as an artist, which was, alas, small. She once sponsored an event at her Greenwich Village studio, according to biographer Barbara Goldsmith, in which each of the other artists was to produce a finished canvas over three days. George Luks, a painter from the Ashcan School, got stinking drunk, then tailed Whitney around the room in a cloud of whiskey. "Mr. Luks, why do you keep following me?" she demanded, finally.

"Mrs. Whitney," he replied, "because you are so goddam rich."

Life is like that for the rich, except within their own enclaves, among their own kind. Mrs. Whitney was still goddam rich when she retreated to her natural habitat on the Upper East Side or to her country place in Old Westbury, but the contrast with her neighbors was not so glaringly awkward. Like her, they had been everywhere, seen everything, perfected the disarming air of being bored with it all. Similarly, it must have been hard for Jeff Bezos to schmooze with the newlyweds in apartment 2-D when his own personal net worth was equal to the gross national product of Iceland. In Medina, his new neighbors can also pass for nation-states.

Arranging the world so that one meets only other rich people thus begins, paradoxically, as an attempt by the rich to live like normal people, like the folks next door.

Then they find that they like it. They stay to cultivate and compare themselves with people of similar stature and to manage the envy of their peers. They find that they are becoming members in a kind of international club, and a peculiar thing happens. You might think immense wealth would free people to become completely different, *sui generis*, themselves. Instead, they typically become more alike. They frequent the same restaurants. They hire the same architects. They buy from the same art dealers in New York and Paris, and if they buy well, they get wooed over time by the same museum curators and auction house specialists striving not to appear too eager. They wear the same kind of clothes. (The shops on the Goëthestrasse in Frankfurt—Chanel, Cartier, Bulgari, Gucci, and so on—are almost identical to the ones on Worth Avenue in Palm Beach or at the Peninsula Hotel in Hong Kong.) They share the same gossip.

The Pseudo-Species

It is all part of the process the Austrian zoologist Konrad Lorenz called "cultural pseudo-speciation," the tendency of human groups to divide themselves into distinct social units almost like species and to create barriers against other groups. This process is of course "immeasurably faster" than the evolution of biological species. It's also more commonplace. Lorenz wrote that "its slight beginnings, the development of mannerisms in a group and discrimination against outsiders not initiated to them, may be seen in any group of children." But he suggested that it takes at least a few generations "to give stability and the character of inviolability to the social norms and rites of a group." When Lorenz was writing in the 1960s, many indigenous tribes seemed to have that stability. The rich, more than most groups, still do. Lorenz, who came from a privileged background, in fact stated his experience of pseudo-speciation in terms of upper-class behavior: "When I meet a man who speaks in the rather snobbish nasal accent of the old Schotten-Gymnasium in Vienna, I cannot help being rather attracted to him; also I am curiously inclined to trust him. . . ."

Behaving like one another, by speaking in the same accents or otherwise, is of course a way for the rich to signal their identity to one another and disarm suspicion: This is one of us, not them.

It's also a way to make subtle distinctions more important. In *The Theory of the Leisure Class,* the University of Chicago economist Thorstein Veblen described the behavior of the rich largely in terms of conspicuous consumption. Yet in many ways *inconspicuous* consumption is more intriguing. Almost all peacocks, for instance, have extravagantly conspicuous tail feathers, which they hold erect and rattle to win the amorous attention of females. The females are hardly indifferent to questions of size and stamina; these qualities, like wealth for the rich, are the price of admission. But beyond that, the females are keenly attentive to inconspicuous details, like glossiness and symmetry in the feathers. If a male loses just 5 of the 150 or so feathers in his tail, picky females tend to avoid his dancing arena.

Among the rich, likewise, inconspicuous signals are a sort of highly nuanced private language for the subspecies. A woman who is a member of the club might for instance wear what appears to be a plain brown sweater. Only her peers would recognize it as Yves St. Laurent silk couture, costing more than, say, her Chanel raincoat. Likewise, at his home in Italy, Sirio Maccioni, owner of the fashionable New York restaurant Le Cirque, drives an unprepossessing Lancia. But a member of the club would know from the throaty burble that it is in fact a Ferrari under the hood. The Agnelli family, whose company manufactures both Lancia and Ferrari, began producing this stealth Ferrari in the 1980s, when leftist politics made it imprudent to display wealth too explicitly.

These signals are often too subtle for outsiders to appreciate, which is at least partly the point: "If you go to a house of someone who's new to Aspen and you see a Cy Twombly, you know they have money, especially when it's a house that's not used very much. It's really saying to those who know, 'I'm really rich.' And people who don't know, you don't care what they think anyway."

That last phrase suggests what Lorenz called the "dark side of pseudo-speciation," the tendency to consider outsiders irrelevant, uninformed, even subhuman. It is an entirely natural tendency. Indigenous groups do it implicitly, as Lorenz pointed out, when they use their word for "man" or "the people" as the name of their tribe, and for nothing else: "From their viewpoint it is not, strictly speaking, cannibalism if they eat fallen warriors of an enemy tribe." And from the viewpoint of the rich?

Encapsulation

The history of wealth has always been about rich people separating themselves from hoi polloi. In different periods and places, the rich have worn clothing denied by law to the lower orders; they have recorded their own genealogy, a useful tool of social dominance, while forbidding poor people to do the same; they have even, at times, spoken a different language (Latin for the medieval gentry, Norman French for the post-Conquest aristocracy in England, Parisian French for the nineteenth-century Russian *grande bourgeoisie* at home, Classical Chinese for the educated and aristocratic in China). The rich have also sometimes gone to extraordinary lengths to avoid the horrific possibility of seeing or being seen by their social inferiors. In the early 1700s, for instance, the Duke of Somerset, one of the wealthiest peers in England, used to travel in his coach-and-six with outriders sent ahead to chase rustics from the fields lest they sully him with their gazing. As recently as 1945, when the Maharani of Baroda went riding on horseback, servants shouted at people on the road to look away.

The result, paradoxically, may have been to reduce envy and competition by people who were not part of the privileged group. "The peasant may not feel deprived compared to the lord so long as the peasant continues to think of the aristocrat as a member of another species, as a different kind of animal with whom comparisons may not be made," writes Jerome Barkow, a Canadian anthropologist. "When such *encapsulation* of social groups takes place, envy is prevented." The rich would never, of course, have suggested that their social inferiors belonged to some distant species; they would have said simply, "Not our kind, dear."

When the rich actually chose to be seen by common people, they tended to elicit awe and subservience, as if they were superhuman. "I like best to visit the Baron in his office," Heinrich Heine wrote about his friend the Paris financier Baron James de Rothschild, "where, as a philosopher, I can observe how people bow and scrape before him. It is a contortion of the spine which the finest acrobat would find difficult to imitate. I saw men double up as if they had touched a Voltaic battery when they approached the Baron." Or as an Egyptian civil servant put it in 1500 B.C., writing on the otherworldly presence of the king: "He is a god by whose dealings one lives, the father and mother of all men, alone by himself, without equal."

But isn't this literally ancient history? The rich no longer live behind moats and castle walls. Bill Gates sits still for questioning by CNN reporters, and Sarah Ferguson, the Duchess of York, will appear, as the British like to say, at the opening of a door.

Yet when I went to visit the rich in their own habitats, it did not seem as though all that much had changed. The "us or them" question always came first. "Where are you staying?" they asked, to which the best possible answer was, "With friends," but only if the friends happened to be part of the club. Failing this, the Little Nell would suffice in Aspen, or Le Crillon in Paris. The rich also sometimes revealed the sense of being separate and without equal: Wealth, George Soros has said, "is a sort of disease when you consider yourself some sort of god, the creator of everything." Then he added: "But I feel comfortable about it now since I began to live it out." Conversely, their servants and staff sometimes confessed to being treated as if they were not quite human. One housekeeper remarked that her employer never spoke to her even when they were in the same room. Instead, she would sometimes phone halfway across the Pacific Ocean to have instructions relayed back by her major domo. The housekeeper meanwhile ghosted a room or two behind her boss, invisibly picking things up and putting them right, always careful, as instructed, to disinfect any doorknob she touched.

I was struck, finally, by how hard it was to reach many of the places where the rich go to find one another. A quality of splendid isolation seemed to be the rule: Aspen is at the top of a narrow valley, and the pass through to the other side of the mountains is shut down all winter. Nantucket, Palm Beach, and Majorca are of course islands, and Lyford Cay is a gated peninsula on an island. San Carlos de Bariloche, the Argentinian ski resort where Ted Turner, Sylvester Stallone, and George Soros have ranches, is in the foothills of the Andes, and at the other end of the hemisphere. The topography of isolation no doubt provides a measure of security. Monaco, for instance, is protected by mountains and the sea, and the police can shut down all access roads in minutes.

Isolation, the sense of being cut off from the everyday world, interested me in at least one other sense. Maybe it was just a coincidence. But these were the very habitats most likely to produce new species in the natural world.

2

The Long Social Climb

From Monkey to Mogul

Descended from apes! My dear, we will hope it is not true. But if it is, let us pray that it will not become generally known.

—ATTRIBUTED TO THE WIFE OF
THE BISHOP OF WORCESTER

EVOLUTION IS AN ASTONISHING THING, CAPABLE, AS ONE SCIENTIST has lately written, of transforming "a tiny, bulgy-eyed, tree-hugging, insect-crunching proto-primate into Julia Roberts." This transformation has taken place almost invisibly, over 70 million years. No single generation in that time, nor even any hundred generations, looked notably different from its immediate predecessors. That's the nature of evolution. It works by endless incremental shifts and the slow piling on of mutations. A few individuals survive a plague because of some happy genetic quirk and their genes proliferate imperceptibly in subsequent generations. One or two male birds develop elongated tail feathers and attract more mates, with the result that long tails eventually predominate. The climate changes, and a few apes swing through the trees, gradually evolving into chimpanzees, while others climb down and wander out onto the grassy plains and become Julia Roberts. To see evolution actually taking place we would need time-lapse animation to pan across the generations in extreme fast-forward.

Or maybe it would be better to view the whole process in rewind. The biologist Richard Dawkins dreamed up a wonderful device to dramatize just how near we are to our evolutionary roots. Dawkins set his scene in East Africa, the site of our actual evolution. For the purposes of this book, Africa is a little too remote, too susceptible to putting our origins at a safe remove from the everyday lives of the rich. So let's imagine, as Dawkins did, that a modern human being is standing face-to-face with a chimpanzee. But let's set the scene in the lobby of The Breakers Hotel in Palm Beach, and let's say the human being is Julia Roberts.

On the Road with Julia Roberts

With one hand, Miss Roberts is holding onto her mother, who in turn is holding onto her mother, and so on back through the generations. The chimp and her ancestors are likewise lined up hand-in-hand, and the two facing lines wind out of the hotel, past the ornate stone fountain of nude women astride spouting dolphins, and left at the end of the driveway onto South County Road. Over on Worth Avenue (we're taking the scenic route here, and why not?), all is well with the world. Miss Roberts's ancestors ogle the $400 beach towels in the window at Hermès and the diamonds at Van Cleef & Arpels. The chimps gaze longingly over their shoulders in search of palm trees. The two parallel lines continue out of town across the Royal Palm Bridge, each studiously indifferent to the other, and north up Route 95 into the distance—but not too far into the distance.

Up around the Georgia border, near the Okefenokee Swamp, something dreadful crops up in Miss Roberts's ancestral line: It's the sort of grunting, armpit-scratching family member seemingly set on this Earth for no better purpose than to mortify the nouveaux riches. Or rather, to mortify new money, old money, and no money alike. In that brief 300-mile span, formed by 350,000 human and prehuman generations standing four-and-a-half feet apart, the line of chimp ancestors and the line of human ancestors have merged, in a single, hairy, chimplike mother of us all.

To put it another way, Miss Roberts could drive past every one of her hominid ancestors in less than five hours. Four if she pushed the speed limit. Genetic evidence indicates that the distance back in time to the common ancestor of chimps and humans alike is roughly 6 million years. This may seem like a while, particularly if you are accustomed to thinking in terms of quarterly results, or when you can pick up next month's trust fund check. But to evolutionists, who date the origins of life in billions of years, it's yesterday.

So the idea of old money doesn't do much for Richard Dawkins. But our proximity to the great hairy mother of us all makes him positively giddy. He imagines walking up and down the human chain "like an inspecting general," contemplating the vanished generations who did not merely stand face-to-face with chimpanzees but actually interbred with them. "Remember the song, 'I've danced with a man, who's danced with a girl, who's danced with the Prince of Wales'?" Dawkins writes. "We can't (quite) interbreed with modern chimpanzees, but we'd need only a handful of intermediate types to be able to sing: 'I've bred with a man, who's bred with a girl, who's bred with a chimpanzee.' "

This is presumably not a tune Julia Roberts will be singing in an upcoming movie. So let's return instead to Route 95 in Florida for a moment, to consider how close we really are to some of the critical developments in hominid evolution. About 4.5 million years ago (that's somewhere north of Daytona Beach in Miss Roberts's line of forebears), an early ancestor dragged herself to her feet and began to walk upright. Bipedalism led in turn to the evolution of enlarged, symmetrical female breasts and the male's prominent penis, as these newly visible areas became a focus for sexual selection. (Erin Brockovich sends her special thanks to all the little people who made this possible, by not reproducing.) Roughly 2.5 million years ago (approaching Melbourne around mid-Florida), early hominids began to use stone tools. At about the same time, the hominid brain began a rapid expansion, culminating about 500,000 years ago (around Jupiter Island) in the modern human mind. With the development of language 150,000 years ago (somewhere in West Palm Beach), Miss Roberts's ancestors began to perform speaking parts, as opposed to just panting and hooting. By the time art first appeared, about 35,000 years ago, the Roberts family had already stepped safely across the bridge and into the land of the anointed. *Ecce* Palm Beach woman.

The First Rich Person on Earth

We could of course travel in the opposite direction, following the line of Miss Roberts's ancestors back 70 million years to that toothy proto-primate. At that point, we'd be somewhere around Skagway, Alaska. So let's not go there, especially as we have already come to a major pothole in the road to a natural history of the rich. Human evolution had arrived at essentially its present form 100,000 years ago, but archaeologists generally believe the first rich person appeared on the scene only about 10,000 years ago.

It started not with Rolex watches or diamonds, of course, but with something like chickpeas. Somewhere around the border of modern-day Syria and Turkey, a tribe settled down and invented agriculture. Archaeologists have generally theorized that population growth, climate change, and food shortages forced this radical shift from our hunter-gatherer past and that a few leading families, the J. R. Simplots of their day, came through some distant winter with a surplus. It may have been einkorn wheat, bitter vetch, or any of the seven or so Neolithic founder crops, but chickpeas loom large in archaeological thinking, and they could easily have provided the first material basis for the whole panoply of behaviors we have come to associate with wealth. Chickpeas could be stored, they could be traded, they could be used to accumulate land, power, sexual partners, and bric-a-brac, roughly in that order.

If evolution was effectively complete 100,000 years ago, and the first rich folk only turned up with their chickpeas 10,000 years ago, what does one have to do with the other? Arguably, nothing at all. Evolution can sometimes work with astonishing speed; but it's been only 400 or 500 human generations since we began to domesticate plants and animals. (Going back to Julia Roberts's ancestral line for a moment, that would be out the driveway of The Breakers Hotel, but not even halfway to Worth Avenue. She could *walk* past all her agricultural ancestors in ten minutes.) In that time, we know of only a few significant changes in human

physiology: Because of dairying, roughly 30 percent of the world population has evolved the ability to digest lactose as adults. This tolerance
occurs mostly in groups with a long history of drinking animal milk;
some scientists date its proliferation back little more than a thousand
years. At the same time, two genetic adaptations to a grain-based diet
have proliferated in populations with a lengthy history of practicing agriculture. (A quick caveat: Natural histories aren't really about genetic differences in any case. It was natural history, not genetics, when the rich
used primogeniture as a way to preserve a family territory, by passing it
on intact to a single heir. It was also natural history when the Napoleonic
Code outlawed primogeniture early in the nineteenth century, and the
rich in Italy responded by marrying their first cousins, for the same purpose of keeping the family territory undivided. Both practices were adaptations to changing ecological conditions.)

On a Perilous Cusp

But isn't it implausible to suggest we could have evolved biological or
even psychological tools for dealing with something as radically new and
different as wealth, with all its lovely comforts (the sudden end of quiet
desperation, the instantaneous satisfaction of almost any whim)? In
many ways, being rich seems like a biological conundrum. It's as if a
cheetah, whose ancestors spent all their time chasing down big game at
sixty miles an hour on the open Serengeti, suddenly found herself in a
zoo with an unlimited supply of fresh meat, and not a wildebeest in sight.
It's as if someone took a wolf, bred for ripping open the bellies of enraged
bison, and turned him, after a few generations, into a bichon frisé pid-
dling on the Serapi carpet at the sight of the UPS man.

It often seems that evolution has left us utterly unprepared for wealth.
Think of King Ludwig II of Bavaria gibbering in his fantasy castles or
Howard Hughes cloistered up with his germ-phobia and his Kleenex.
Early in my research among the rich, it seemed to me that even the sane
ones often had a special relationship to reality. They suffered from desperate acquisitiveness, and they were often compulsive about secrecy
from the outside world. Author Connie Bruck recounts how junk bond

financier Michael Milken was once on the phone with investor Nelson Peltz when Peltz's pet parrot started squawking.

"Are we alone?" Milken asked.

"Yes."

"What's that noise?"

"A parrot," Peltz replied.

"Call me back when the parrot's gone," Milken said, and hung up.

Adjusting to unimaginable wealth is difficult for anyone. Even the Rothschilds complained about "living like drunkards" during the years of their first success, never certain if they were millionaires or bankrupts. And five or six generations of practice do not always seem to help. In 1996, two hundred years of du Pont breeding and tradition reached their full flower in the person of John E. du Pont, who sometimes called himself "the golden eagle of America" or "the Dalai Lama of the West." He also liked to drive around Foxcatcher, his 800-acre estate outside Philadelphia, in a tank. Du Pont subsidized the U.S. Olympic wrestling team, indulged fantasies that he was himself a champion athlete, and ended up murdering his team's coach. He is currently serving a life sentence in prison. On the other hand, a friend of one of my friends belongs to the peculiar breed called New England Old Money and lives in almost total denial of her wealth. She inherited five homes from her mother, and caretakers maintain them for her, but she lives modestly with her significant other, a woman from South America, who makes a living cleaning houses.

An understandable conclusion would be that wealth is unnatural, and the behavior of the rich often downright maladaptive—that is, tending more toward extinction than to evolutionary triumph. The rich, one anti-Darwinian writer has argued a little too comprehensively, suffer from "a proclivity . . . to early sexual exhaustion, to sexual incapacity . . . to homosexuality, to religion . . . to art, to connoisseurship . . . to almost anything in the world, in fact, except increasing or even maintaining the numbers of their own class by reproduction." Reading this, I imagined every new rich person in the same spot as those first agricultural plutocrats 10,000 years ago, trying to figure out how to remake their lives, standing with their chickpeas on a perilous cusp somewhere between the Pleistocene and the epicene.

But I also knew it was hogwash. We focus on the failures and flameouts among the rich in part because of our own highly developed sense of schadenfreude. If we cannot be rich ourselves, it at least consoles us to imagine that the rich are unfit for the job and unhappy once they get it. This was the spirit in which Dorothy Parker once remarked, "If you want to know what God thinks of money, just look at the people he gave it to." But in the course of my research I met plenty of rich people who were relatively normal. Most of them had overcome the sleepy demon of sexual incapacity at least long enough to reproduce themselves. Indeed, I was impressed with how many wealthy families had found the delicate balancing point between reproductive insufficiency and Malthusian excess, such that they had maintained their wealth and position more or less quietly for generations: The Goelets, the Fords, and of course the Rockefellers in the United States; the Grosvenors in England; the Rothschilds in England and France; the Porsche/Piech and Haniel dynasties in Germany; the Agnellis in Italy; the Mitsui family in Japan. If the du Ponts have produced the occasional misfit like "Dalai Lama" John, it's surely more significant that they have also produced Winterthur, an extraordinarily tasteful display of, well, art and connoisseurship, not to mention a very large corporation, DuPont, and a very small state, Delaware, all of which remain under their considerable influence. The rich aren't by and large madmen or incompetents drifting haplessly in some evolutionary lacuna. In Darwinian terms, they seemed to me to be more like the winners.

Rich People Live Longer

Consider, for example, the question of dying, which seems at least on the surface to be a reliable indicator of having lost the Darwinian struggle. Rich people die, too, of course—just not so soon. They lead longer, healthier lives than the rest of us. The old cliché holds that all the money in the world doesn't mean a thing if you don't have your health, but people with money usually do. And the more money they have, on average, the better their health. The 1990 Longitudinal Study in the United Kingdom found that homeowners with one car tend to die younger than homeowners with two cars, and so on in a "continuous gradient" of decreasing mortality from the most deprived areas to the most affluent.

(The study merely took car ownership as a handy measure of affluence; it did not mean to imply that having twenty cars qualified Elton John for immortality.)

Other research indicates that wealthy people also lived longer in the past. The word "wealth" itself suggests as much. It comes from "weal," which means "well-being." In one of the stranger pieces of demographic research on record, a team of epidemiologists and psychologists prowled the cemeteries of Glasgow in the mid-1990s armed with chimney sweep rods. They used them to measure the height of more than eight hundred nineteenth-century obelisks. People buried under obelisks tend to be affluent, and the researchers assumed that taller obelisks marked the graves of the more affluent people. The study revealed that every extra meter in the height of an obelisk translated into almost two years of additional longevity for the people buried underneath.

Likewise in Providence, Rhode Island, in 1865, the taxpayers, comprising the more affluent members of society, had less than half the annual death rate of nontaxpayers. In fifteenth-century Florence, fathers who made the largest dowry investments for their daughters in an endowment fund called the *Monte delle doti* had half the annual mortality of the somewhat less affluent fathers who made the smallest investments. If they cannot take their wealth with them, the rich do at least seem to use it to delay the trip. In Los Angeles not long ago, a self-made centimillionaire showed me a golden hourglass, filled with liquid, which he gave himself for his fiftieth birthday. It marked the passage of time in intervals not of three minutes but of thirteen, and not with grains of sand but with a plume of diamonds.

A Sense of Well-Being

How does wealth translate into longevity? Money of course buys better medical care. So the connection between wealth and survival might seem unsurprising at first. A multimillionaire lawyer I met in Palm Beach actually dismissed one of the great philanthropic passions of the rich—supporting hospitals and naming new wings for themselves and their kin—as little more than a Darwinian bid to ensure the best possible care in the ultimate arena for survival. John D. Rockefeller, for instance, produced heroic benefits for humanity at large when he created one of the world's great medical research facilities, the present-day Rockefeller University. He also reserved four private rooms in the first sixty-bed hospital exclusively for his family. This sort of bid for special treatment can work wonders: J. Seward Johnson, of the Johnson & Johnson fortune, was once a patient at the Medical Center at Princeton, and the administrator and chief fund-raiser personally chauffeured him home and back when Johnson felt this might bring relief for his constipation, a service not covered under the average person's medical plan.

What's more intriguing about the finely graded link between health and money is that it defies such straightforward economic reasoning. "It looks as if what matters about our physical circumstances is not what they are in themselves," one demographer concludes, "but where they stand in the scale of things in our society." Clearly, it is best to stand on top. Rich people by and large enjoy reduced stress, greater social support, and a distinct sense of personal control. It may not seem this way to some third-generation heiress suffering rage and frustration at the limits imposed by her paternalistic trust fund officer, but there is a subliminal serenity in knowing one's rent will be paid next month, and for the 653 months after that, too.

The rich get to look around their foursome, on a particularly fabulous day out at Pebble Beach, and enjoy the warm sense of well-being that comes from knowing they're just a little bit more fabulous than their partners. The habit of being in control seems to cushion the rich from time's ravages. When estate taxes are about to be reduced, people with money apparently "will themselves to survive a bit longer" to get the benefit, according to a recent University of Michigan report. They are dominant animals, and biological studies of rats, guinea pigs, and other species indicate that dominants go on living their charmed lives while subordinates drop dead around them. John D. Rockefeller lived to be ninety-eight, and biographer Ron Chernow offers the redeeming note that he never in fact used those private rooms at his hospital. This is precisely the point: He didn't need to.

Living longer doesn't necessarily translate into Darwinian fitness. (There may actually be a trade-off in humans between how long we live and how many offspring we produce, the standard measure of Darwinian success.) But it's what evolutionists speak of as a proximate factor. That is, it's what we really want, just as sex is what we want, approximately now, and babies are often what we get, a little later, as an unintended result. We drink our toasts, "*L'chaim!*", to life, and a little less enthusiastically to the joy of having twenty-three grandchildren.

Our Fierce and Elegant Kin

So let's assume that evolution has in fact prepared the rich quite well for their good fortune and that we can learn about their success by looking at antecedents in animal behavior. Where do we begin? Larry Ellison, founder of Oracle Corporation, is not a chimpanzee, nor is Nicky Hilton, heiress to the hotel fortune, a bonobo. But it's a near thing. Chimps and bonobos are our two closest relatives on the planet, and to understand how certain behaviors of the rich developed, it will help to know just how close we really are.

This requires some brief taxonomic background: Within the scientific order *Primates*, there are three separate superfamilies—the prosimians, monkeys, and apes. The apes, or *Hominoidea*, include gorillas, orangutans, chimpanzees, bonobos, and humans. Most textbooks then split humans off into our own family *Hominidae*, in which *Homo* is the only genus and *sapiens* the only species. But biochemical evidence indicates that chimps and bonobos are *Hominidae*, too. These two African apes are so similar to each other that scientists did not recognize them as separate species until 1929. Bonobos are about the same size as chimps, though they have longer hair and a more "gracile and elegant" build, according to primatologist Frans de Waal, who adds, "Even chimpanzees would have to admit that bonobos have more style." The two species differ in just 0.7 percent of their DNA. Humans differ from each of them by only about twice as much, 1.6 percent of our genome. The physiologist Jared Diamond notes that the gulf between us and them is considerably narrower than the 2.9 percent difference between red-eyed vireos and white-eyed vireos, two bird species the average layman might lump together as so much olive-green fluff. "The remaining 98.4 percent of our DNA is just normal chimp DNA," writes Diamond.

The inference the reader is liable to draw is that 98.4 percent of our behavior is just normal chimp behavior. To play devil's advocate for a moment, there's at least one other way to look at these numbers: If nature is a bit of a cheapskate with her raw materials, she is also a procreative genius, adept at spinning even a small difference in the genome into vast differences between species. Genetically, chimps and bonobos are 99.3 percent identical to each other. Behaviorally, on the other hand, they are almost opposites. Chimps live in groups dominated by brutal males, who bully their females, engage in dull, repetitive sex, and delight in leading murderous "party gang" raids on their neighbors. Bonobos, on the other hand, live like voluptuaries in an opium den, and their female-dominated society relies on friendly, inventive sex to keep everybody relaxed. There can be tension when rival groups meet, but it soon subsides into mass copulation. "The chimpanzee resolves sexual issues with power," de Waal writes, "the bonobo resolves power issues with sex."

War Parties or Mass Copulation?

So where do rich people fit in? Humans seem to have it both ways: We live for the most part in male-dominated societies, and we practice "party gang" violence on a scale otherwise unknown in the animal world. When *The Sunday Times* of London recently put together a list of the two hundred richest Britons since William the Conqueror, half of the top ten had made their fortunes in war or conquest. It's also easy to see a chimp model in the corporate raiding parties of John D. Rockefeller and his male cohort in the late nineteenth century or of Michael Milken and his cohort in the twentieth. But we also rely heavily on gentler forms of social manipulation, including strategic sex. On the bonobo side of the ledger, negotiations that freed India from British rule may have gone more smoothly because Jawaharlal Nehru, leader of the Indian side, and Lady Mountbatten, whose husband represented the Empire as British Viceroy, were using sex to resolve issues of power. (Or was it the other way around?)

Some scientists would caution against extrapolating from the behavior of either species. Our resemblance to chimps, bonobos, and other primates is in some ways as close as an echo. In other ways, it's as if the echo of a grunt has somehow come back transformed into a beautiful aria. We have language; studies of so-called "ape language" have served mostly to demonstrate that other species don't. We have a theory of mind, a biological term meaning that we can recognize the thoughts and feelings of other individuals; chimps, bonobos, human children under the age of four (and, OK, some rich people) don't, or possess at best only the vaguest, inchoate awareness of other minds. Above all, we have the collective knowledge and custom of civilization; chimps have only cultures, localized traditions in the use of tools and foods.

All these astonishing differences are recent events in human evolution. If we were to trace their development in Miss Roberts's line of ancestors, we probably wouldn't get much farther than Jupiter Island, or roughly 500,000 years ago. In the evolutionary scheme of things, these differences are not all that much older than the advent of agriculture and human wealth. We should not be so astonished by them that it blinds us to all that we have in common with other primates. While some of the behaviors of the rich and the fashionable would doubtless send the average primate brachiating desperately for the nearest exit, others would be deeply familiar.

Mwah-mwah

The air kiss, for instance, may seem like a modern affectation, "a snobby kind of mannerism," as the etiquette expert Letitia Baldrige has put it. But in truth there's something ancient going on when fashionable partygoers greet one another by pooching out their mouths, parting their lips with a moist clicking or chirping sound, and saying *mwah-mwah* into the air beside one another's ears. Primatologists call it lip-smacking. It's a ritualized imitation of grooming behavior. Monkeys and apes use it as a gesture of appeasement to disarm potential rivals. Like partygoers, they also sometimes use their version of the air kiss deceptively, as a prelude to backstabbing. For instance: The American biologists Robert Seyfarth and

Dorothy Cheney lived for years among a troop of vervet monkeys in Kenya's Amboseli National Park. One day they observed a female being groomed by a male named Escoffier. The female's daughter, a vervet princess named Leslie, apparently disapproved of this relationship and launched an assault on Escoffier, who fled. Leslie plunked herself down and groomed her mother for a few minutes, proprietarily. Then she walked over to Escoffier, who cowered. The remorseful Leslie lip-smacked, and Escoffier allowed her to approach and groom him, a case of air kiss and make up, or so it seemed. Leslie swept her fingers through his hair, picking out burrs, dirt, invisible bits of dander. Escoffier gradually unwound. Finally he stretched himself out languidly, so Leslie could groom his back. "At this point," the biologists write, "Leslie picked up Escoffier's tail and bit it, holding it in her teeth while Escoffier screamed."

The incident suggests that some behaviors have indeed passed intact from our monkey forebears directly to the *Forbes* 400. We still lip-smack exactly like monkeys. And having disarmed our rivals, we also still bite them on the ass, though nowadays it may not happen until three days later, in the form of a blind item in the gossip columns.

Behaviors into Language

The unique medium of human language is responsible for some of the most dramatic transformations of primate behavior. For instance, monkeys spend hours every day grooming one another; it's their major tool for social bonding. Humans, being primates, also have a primordial need for stroking, but we've learned to do the same thing with chitchat and particularly, as we shall see later, with flattery toward our social superiors. This kind of transformation of behavior into words makes sense. The average primate has no more than twenty distinct calls; the average English-speaking human has sixty thousand words. Language undoubtedly also helped produce a transformation that would prove critical in the rise of wealth on this planet: Chimps and some other primates share food and practice social manipulation through raucous feeding clusters. Humans, particularly the rich and ambitious ones, do the same thing with feasts and parties.

In fact, the overriding impression a human being takes away after sitting down for a few days among other primates is how closely their social lives resemble ours. The part of our heritage that drives us to seek status, to put some people down and build others up, to gossip, to form social alliances, and to build on family connections is vividly on display anywhere two or three monkeys gather together. Social life is the great area of primate specialization, one of the things that unite us as a group. Proponents of the so-called Machiavellian Intelligence hypothesis argue that our social nature may even have been the critical factor in the evolution of the enlarged primate brain.

Thirty Million Years of Social Climbing

"Social intelligence is for primates what celestial navigation is for arctic terns," Robert Seyfarth said, when I visited him and Dorothy Cheney in Botswana's Okavango Delta, where they now study baboons. This intelligence isn't exclusive to humans, or even to our cousins the bonobos and chimps. In fact, roughly the same social system is present in so many of the three hundred modern primate species that it had probably already reached an advanced stage of development by the time the ape lineage, our own *Hominoidea,* separated from the mere monkeys, roughly 30 million years ago. That is, we were status seekers and social climbers for millions of years before we were even human. We were practicing to be rich while we were still swinging in the treetops.

One morning with Seyfarth and Cheney, I stopped at a gathering place I called The Palm, where a couple of baboons were sitting around taking in the sun, being groomed by subordinates, and waiting for breakfast to fall out of the sky. A baboon was eating palm nuts in peace at the top of a tree and inadvertently shook a few nuts down to the ground. Power and Selo, the king and queen of Troop C, took the first couple of nuts. When they moved off, Sonny, the number two male, slipped into their place. Sonny threatened a lesser male named Gary with a sharp lifting of the eyebrows. Gary immediately recouped his pride by passing the threat on down the line to a hapless juvenile. It was a behavior biologists call redirected aggression, and of course we all do it, kicking the dog being the dismal last resort.

Looking at this power breakfast, I began, with a rather plodding literal-mindedness, to tick off many of the basic behavioral mechanisms ascribed by evolutionary biologists to humans and animals alike. Over there, a bit of reciprocal altruism in the way one baboon finished grooming another, then closed his eyes and tilted his head to be groomed in turn. Over here, a little kin selection in the way a big brother rushed to discourage some lout from troubling his sister. And there, dominance behavior in the way one baboon supplanted another at some choice spot in the sun.

"It's like sitting in an Italian café nursing a glass of wine, and seeing the teenagers and the young adults coming and going," said Seyfarth, "each with a complex story." At times someone slipped away to avoid running into a rival. Other times two friends met and strengthened their bonds by sitting together. Cheney and Seyfarth didn't just watch the comings and goings. They also spread false information. Their technique was to use playbacks of the baboons' own vocalizations to determine just what the baboons themselves knew about the complex social connections within the troop.

Who's Who?

It turned out that they knew a lot, not just about their own circle of family and friends, but also about the family and friends of the fifty or sixty other individuals in the troop. If the researchers played the scream of a baboon named Bridget, for instance, her mother, Jane, naturally took note. But other monkeys also recognized Bridget's voice and looked to Jane. If Jane happened to be sitting with a baboon named Hilary, who outranked her, the researchers might play a fight sequence between two of their relatives. Hilary and Jane typically responded by looking in the direction of the fight. Then they looked at each other. In the next fifteen minutes, more often than not, Hilary would do something to drive Jane from her seat. Jane was unlikely to take revenge on Hilary, because baboons try not to pick fights they can't win. She'd go off instead and supplant Hilary's best friend—more redirected aggression—or threaten one of Hilary's children.

The baboons were as acutely alert to who's who in the troop as Milanese fashion mavens racing through this morning's gossip columns in search of boldface names. They were as sensitive to nuances of rank and family relationship as the ambitious young Deborah Mitford browsing through *Burke's Peerage* for potential husbands, earls or better only, preferably not too repulsive (she ended up with Duke of Devonshire). Using this laser beam intelligence to manage complex social relations is something primates have done for eons.

Indeed, sitting among the baboons, I got the unsettling impression that social relations mattered more to them than food or sex. Food was everywhere, and sex was largely the privilege of dominant individuals. But questions of status, or prestige, made all heads snap to attention. This was unsettling because food and sex are what the Darwinian struggle is all about. But the socially prominent couple of the moment weren't even having sex, though Selo was in the full red-rumped blossom of sexual readiness. They were in fact a dysfunctional family; Power had grown up with Selo and, though she was not his sister, he was apparently suffering inhibitions against incest. Yet the two of them paraded around together, accepting the accolades of their social inferiors, as if all were right with the world. Power wasn't letting any other male get near Selo. She was his arm candy.

In a deeply political species like the chimpanzee, the similarities to the social behaviors of the rich become even more pronounced. Chimpanzees are masters of social networking, with a Machiavellian knack for developing friendships and building political alliances. Prominent individuals practice a kind of *noblesse oblige*; they seem to understand that sharing food and other resources is a way to accumulate prestige and the support of lower-ranking individuals. Like the rich, chimpanzees also know the value of putting on the right face. In his book *Chimpanzee Politics,* Frans de Waal recounts the spectacle of a dominant male named Luit being challenged from behind by a rival. Before turning to meet his challenger, Luit paused, like a CEO about to enter a roomful of dissident stockholders, and actually reached up with his fingers to press his lips together and wipe away his nervous grin. Then he faced down his rival with the serene image of unshakable power.

Wah-Hoo Contests

One day during my visit with Troop C, the stillness of the Okavango Delta suddenly erupted, and a ninety-pound male came hurtling past, hackles raised, canines bared, screaming with all his heart at some rival vanishing just ahead in a tumult of dust. Everybody else in the troop scrambled for safety. The chase led up into the treetops, where the two thugs sat on branches roaring "Wah-Hoo! Wah-Hoo!" at each other, a sound like a hurricane being cooped up in a bottle and let out in brief, deafening bursts. These competitions between prominent males, called wah-hoo contests, sometimes went on at considerable length, but they hardly ever progressed to real violence. Baboons and many other species practice ritualization; that is, they use small gestures derived from some larger pattern of movements to symbolize their intent. This is particularly true in contests for dominance. For them, as for us, thrusting the head forward and dropping the eyebrows in a sharp glower can be just as effective as (and considerably safer than) actually throwing a punch. Likewise, red deer engage in roaring contests; the winner isn't the one who gores the other in the underbelly, but the one who merely demonstrates his ability to do so by roaring loudest and longest.

The analogy to our own battles of the titans is perhaps a little too obvious: When fashion designer Calvin Klein calls clothing manufacturer Linda Wachner a "cancer" on his brand, or when the late Helena Rubinstein leaned out the window of her Fifth Avenue office to scream at cosmetics rival Charles Revson, "*Vat* are you doing? You're killing me, you rat!", or when Canadian publisher Conrad Black describes a politician as "a repulsive little gnome—greasy, twitchy, and specious," we are deep in the land of the wah-hoo contest.

Looking around the baboons of Troop C, it seemed to me that many of the basic elements of a natural history of the rich—the social climbing, the backbiting, the elite of alpha males and females who set the terms of daily life for their social inferiors—were already in place 30 million years ago. The only thing missing was, of course, the wealth.

No other animal has anything like it. No other primate does anything even as simple as storing food for the off-season. So how did we get from Troop C, stuffing their cheeks with jackalberries and picking one another's ectoparasites, to the socialites sweeping into The Breakers Hotel for the annual Red Cross Ball, or the moguls eating warm sweetbread salad at Le Cirque?

It took agriculture, of course. A more intriguing argument is that it may also have taken a few really good parties.

3

Party Time

The Dawn of the Plutocrat

What is your hosts' purpose in having a party? Surely not for you
to enjoy yourself; if that were their sole purpose, they'd have simply
sent champagne and women over to your place by taxi.

—P. J. O'ROURKE

AT THE BOTTOM OF RUTHIE'S RUN ON AJAX IN ASPEN, SKIERS
sometimes pause to watch the afternoon light washing down the raw
slopes of Red Mountain, across the valley. Red Mountain itself is unpic-
turesque, flat-topped, hulking. So they're only likely to soak up the splendor of it all if they actually happen to own a house there. The steep
southern flank of the mountain is a showcase for the grandiose homes of
the newly rich, homes not meant to be approached too closely by the
uninvited (all roads on Red Mountain are private) but to be seen and
admired from a mile or two away. Peak House, for instance, is what a cozy
Tyrolean hill house might look like if it were blown up to 24,000 square
feet and built on a fifty-five-degree grade: Overhanging slate roofs held up
by ornately carved and corbeled timber brackets, with a huge patio and
pool in front, atop an 18-car garage. Downhill, on a beautiful property that
was once a ranch, Leslie Wexner, boss of Victoria's Secret and The Limited,

has a house big enough to need a bank of furnaces, many of them for melting the snow on the driveway. He also has a tepee in the side yard, a common sort of reverse status symbol in Aspen, to show that one is at heart a simple soul, in touch with the great spirit of the mountains.

The houses on Red Mountain are of course impressive inside, too. Sometimes seriously impressive: In the living room of one house, the huge wall of glass looks back at Ajax and its neighboring ski slopes as if it owns them. There's a Steinway grand, and a low black granite bar with a well-thumbed coffee table book about the abstract expressionist Hans Hofmann, whose paintings hang on the wall. An Alexander Calder mobile, black and unobtrusive, wings silently through the room like a flock of geese.

Other interiors are impressive, but not so serious. Around the corner in the Starwood neighborhood stands a modernistic, copper-roofed house, which looks, according to locals, "like spaceships colliding." It's equipped with his-and-hers lap pools because she likes the water a degree or two warmer than he does. There's also a disco with a smoke machine for posing and smoldering, and a ballroom for when the mood is more romantic. The pocket doors to the master bedroom shimmer with the golden image of Gustav Klimt's "Kiss," and when you cross an electronic beam the kissing couple splits in two, then reunites, as the doors automatically open and close again behind you. The bed itself is round with faux wisteria overhead. Push the right buttons, and the curved wall of windows rolls away, allowing passionate lovers tangled in the sheets to glide out under the stars in an erotic apotheosis of yin and yang, sturm und drang, climax and anticlimax (the stars, of course), preferably with "Thus Spake Zarathustra" booming in the background, though Neil Diamond singing "Longfellow Serenade" might do in a pinch.

The Munificent Mr. Kozeny

Into this surreal world in the spring of 1997 came a young Czech couple with a penchant for traveling in separate bulletproof cars. They took one look at Peak House and paid $19.7 million cash for it. Heads perked up all over town. Viktor and Ludka Kozeny, the proud new homeowners, promptly set out to do what rich people usually do when they want to establish their place in the world: They threw a party.

Not an ordinary party, either, but one that would be memorable even by the outlandish standards of Aspen at Christmas. Viktor Kozeny was what anthropologists call an aggrandizer, or "Triple-A personality," for "acquisitive, aggressive, and accumulative." He'd made his name in the early 1990s buying up privatization vouchers offered to every citizen in post-Communist Czechoslovakia, then converting them into ownership of formerly state-run industries. At first the mutual fund company he created prospered, and Kozeny became a national celebrity. When it faltered, and the government began to investigate his methods, he found it prudent to leave his native country. With a fortune estimated at between $200 and $700 million, Kozeny had set up house on Lyford Cay in the Bahamas, where *Fortune* reported that he spent $14 million on the swimming pool alone. A few months after buying the place in Aspen, he paid another $25 million for a house in London's Belgravia district. Even in the thick of this house-buying spree, Kozeny believed in the value of living large: One night he managed to spend $21,000 on dinner for four at Le Gavroche in London—sipping one glass from an $8,300 burgundy, Romanée-Conti 1985, and sending it away for the staff to drink because it was "too young." In Aspen in December, when passion fruit is out of season, he required a glass of fresh-squeezed passion fruit juice at breakfast each morning. An eight-ounce glass took an entire case of the fruit and cost $120. He had a habit of ordering exotic foods, a bird's nest for soup, for instance, "and he'd want it the next day," according to a former staffer, "and there's really only once place in the world that they come from, where these Chinamen have to climb up a cliff and steal the bird's nest. Quite sad really." Whatever Kozeny wanted, from any place in the world, it was on the plane next day into Aspen.

So when Kozeny set out to host his new neighbors at a Christmas party on Red Mountain, expense was not an issue, except that "too much" was always preferable to "just right." Kozeny tried to hire Elton John to provide the entertainment and wound up with Natalie Cole. Word got around. A friend who had houses both on Red Mountain and in the Bahamas had been touting Kozeny around Aspen. People clamored for invitations, even calling around to see whose husband or wife would be out of town so they could go as a date. Between Friday afternoon and Monday, the day of the party, the guest list doubled to 150 people.

At the top of the entry stairs that night, the party planner whispered in Kozeny's ear about each arrival, "so he could act like they were his newest best friends." Then a server presented fresh-squeezed tropical fruit drinks in hollowed out baby pineapples and pomegranates. Inside, the Iranian Beluga caviar was slathered on with a silver trowel, and the shaved white Alba truffles were served like potato chips. Cristal and Château Mouton-Rothschild flowed in rivers. Most of the guests were rich or beautiful, often both, and they had seen everything, but something about the party flipped a switch in many of them. Maybe it was the decorative extravagance, or the lavish menu with six separate entrées, or the chic party favors from Asprey & Garrard. They raced to see how much expensive food and wine they could consume. Swank young women trolled the crowd for wealthy men inclined to vigorous intellectual conversation. (As the home of the Aspen Institute, Aspen prides itself on weighty pillow talk). When the event photographer passed by, they struck exotic poses. It was, one partygoer later recalled, like something out of Fellini. And the center of attention was, of course, the munificent Viktor Kozeny.

Partying to the Top

The usefulness of a good party is, as it happens, one of the oldest ideas in the history of human civilization. Anthropologists regard feasts as a leading means by which the social elite have traditionally gained status and managed social obligations. According to one theory, parties, or competitive feasts, may even have been the driving force by which the rich got rich in the first place.

Academics unfortunately do not focus on modern socialites throwing rival parties and vying to the death for A-list guests. They prefer to model their ideas on tribal feasts, like the potlatches of the Kwakiutl and other Kwakwaka'wakh Indians in the American Northwest. But the similarities are striking. The focus is almost always on how lavishly the host can spend, even if it means going into debt for years to come, and how much he is prepared to give away. Everything is carefully orchestrated to advertise prosperity and the benefits of affiliation and generally to secure the host's place in the leadership of the community.

These highly ritualized feasts occur, when times are good, in tribes worldwide. Because better parties often translate into greater social status, the feasts typically become competitive. Every host wants to do something a little different from the last party, a little special, something people will talk about and remember. A tribal chieftain might wow the crowd with a nine-foot-long yam. A member of the modern urban tribes needs to reach a little further: On New Year's Eve 1999, for instance, two New York chieftains and their wives welcomed 250 friends to a $1.1 million party at Windows On The World atop the World Trade Center. The cocktail hour featured a light show in which the famous faces of the dying millennium were projected onto mimes sheathed head-to-toe in white. One of the male hosts had his own inventive head projected onto Thomas Edison's body; the other presented himself as Copernicus. When the dazzled guests went in to dinner, the blackout screens were down on the view for which the restaurant was famous. Instead, the focus was on the tables, each with two hundred roses arranged in different compositions and dusted with gold. The emcee announced that the hosts had something special they wanted to give their guests, and then a new light show began, with images of the rise of Manhattan projected on sails artistically draped around the room. The music rose to a crescendo over two or three minutes and finally the blackout shades swept up like a theater curtain to reveal the glittering skyline as Frank Sinatra sang, "I'll Take Manhattan." A little gift, something of ours we want to share with you.

The Party as Ritualized Warfare

The party is the defining act in the lives of the elite, and the ambition is typically not just to astonish but to obligate. Whether the partygoers happen to be big men in Manhattan, Kwakiutl chieftains, or Triple-A aggrandizers in the Levant 10,000 years ago, such a party imposes an unspoken burden to repay the hospitality. A little reciprocal altruism six months down the line, a day or two of your time, an important introduction, a modest investment. Lavish entertainments, a Rothschild brother once remarked, are "as good as bribes." For the Kwakiutl, an individual defaulting on the implied social obligation lost not just prestige but the help of important allies, and the chance for advantageous marriages. If the breach led to warfare, he could also lose his life. Modern rich people tend to be subtler than that, using the cold shoulder or the hostile takeover in lieu of the stone ax, but their parties remain strategic at heart. Some academics argue that as an instrument for winning and retaining social status, the feast was a ritualized proxy for war. In the thick of the social season in, say, Palm Beach, it can feel that way even now.

The idea that the competitive warfare of the feast may have been the means by which humans achieved real wealth in the first place comes from Brian Hayden, an archaeologist at Simon Fraser University in British Columbia. Agriculture didn't invent the rich, Hayden argues. On the contrary, *the rich invented agriculture*, and they were motivated by the urge to find that extra little something for a really memorable feast.

Hayden first made this happy suggestion in the late 1980s, in an iconoclastic paper titled "Nimrods, Piscators, Pluckers, and Planters: The Emergence of Food Production." So skeptical readers may hear the tintinnabulation of greed is good Reaganomics trickling down into the world of ethnoarchaeology. But Hayden's argument was intriguing.

The Old Guard Passes

Throughout our evolution, humans had always been hunter-gatherers, and *generalized* hunter-gatherers at that: We lived in small tribes and wandered the countryside scrounging a living from seasonal foods, much like a troop of baboons traveling from jackalberry to fig tree and back again.

The Old Guard of generalized hunter-gatherers appears to have had an egalitarian society based on the sharing of resources. This isn't to say that they were savage innocents; they undoubtedly vied for rank and status. Even the dreariest of these tribes must have included what Hayden calls "people with extreme self-interest characteristics," who yearned for something just a little bit better, like Viktor Kozeny growing up in dull, gray Warsaw Pact Prague. But generalized hunter-gatherers had no real treasures to hoard, nor any good means of hoarding them. Lugging a jewelry box from hunting ground to berry patch would have been deeply impractical. Romanée-Conti was not even a faint bouquet on the horizon, and the idea of locking it up in a cellar to ripen for another ten years would have been anathema.

Things began to change with the appearance of *intensive* hunter-gatherers roughly 25,000 years ago. This new breed learned to exploit more abundant plants and animals, notably grass seeds and fish, with the help of new technologies like basketry, netting, harpoons, snares, the bow and arrow, the grinding stone, and the fish hook. Intensive hunter-gatherers could live in one spot much longer and support a larger population, meaning a more elaborate social hierarchy. They began to see their place in the world differently. They started to store food and accumulate wealth, a revolutionary moment in the history of primates.

The Hoard

Hoarding is a relatively common behavior in other animal groups. Honeybees and ants work diligently to build up a food surplus, and so we use them to inculcate in children the moral value of working hard, saving money, and getting ahead. But ants are Communists. They do their hoarding on strictly egalitarian lines. The practice of agriculture, which also exists in the ant world, doesn't by itself change this. Leaf-cutter ants, for example, are excellent farmers, gathering up leaves from the tropical rain forest, carting them back to the nest, and using them to grow and harvest edible fungus. But like other ants, they continue to practice trophallaxis, meaning that they regurgitate food from their crops into the mouths of their nestmates, until all their comrades are equally full. So ants are probably not the best model in the natural world for human wealth-building. It is true that wealthy partygoers in Ancient Rome used to engage in bouts of bingeing and purging, but we have no record that they ever used their fellow guests as *vomitoria*.

To understand our own hoarding behavior, we might do better to look at squirrels, chipmunks, and other species in which individuals store food for their own private use. Like wealthy humans, these individuals often stock up far more wealth than they are likely to need. The European mole, described by one writer as "a raging sociopath of the countryside," actually stores living animals in her larder. She nips the front end off an earthworm, which apparently renders it comatose. Then she carries the worm home, ties it in a knot, and tucks it into a little chamber in the gallery wall of her nest mound. In one mole's well-stocked house, a biologist once found 1,280 live worms, weighing more than four pounds. Like a connoisseur toddling down to the cellar for an Armagnac 1900, the well-prepared mole can retrieve a healthy earthworm and gobble it up to brighten an otherwise dreary winter evening. But relentless accumulation in the animal world may ultimately be a matter of life or death. Alaskan squirrels cache up to 16,000 white spruce cones each year, and they may need several years' surplus to survive a bitter winter.

A Bias for Sharing?

The trouble with all of these examples is that they don't involve primates, because in fact no other primate on Earth practices anything remotely like hoarding. Monkeys and apes in the wild may occasionally try to hide some choice morsel or stuff it down before anyone else notices. But it's bad form, old chap. Seeds and fruit, their usual diet, are generally abundant and widely distributed. Even when chimpanzees hunt meat, at considerable risk to the hunters' own well-being, the successful hunters typically let out a distinctive call to attract other chimpanzees. The hunters bring their catch to the ground, where a feeding cluster forms, with other chimps all around them begging and whimpering for scraps.

Human hunter-gatherers do much the same thing. Among the Indians she studied in the Brazilian rain forest, the American anthropologist and primatologist Katharine Milton writes, "Individuals do not amass surplus . . . no hunter fortunate enough to kill a large game animal assumes that all this food is his or belongs only to his immediate family." A big kill is an occasion for the whole tribe to gather together for a party—a feeding cluster.

For chimps and hunter-gatherers, food-sharing is a way to gain and keep status. In one study of chimps in the Mahale Mountains of Tanzania, for instance, an alpha male named Ntologi was adept at using the feeding cluster as a kind of bribe. Ntologi usually fed females, influential older males, and unthreatening midrank males; he seldom wasted his efforts on juvenile males, and he never invited the beta male who was his chief rival. Frans de Waal writes, "Ntologi's tenure at the top of the Mahale community lasted an exceptionally long time—more than a decade. Perhaps the way he distributed meat was part of the secret." Sometimes Ntologi simply held onto a carcass and let others pull off meat without bothering to take any himself, further enhancing his prestige by his evident selflessness. Curiously, the Rothschild brothers employed much the same strategy in their lavish entertainments during the first half of the nineteenth century: They provided their guests with the best French cooking, Niall Ferguson writes in his history, *The House of Rothschild*, "though they themselves never tasted a mouthful of it."

In humans and chimps alike, the rewards of being the host at a feeding cluster are political, economic, and often sexual. In a hunter-gatherer culture in Paraguay, for instance, sharing the kill seemed to help successful hunters attract a disproportionate amount of extramarital sex. The female anthropologists who studied this culture suggested that women have sex with successful hunters to encourage them to stay with the group. But the idea of having sex for the good of the group somehow rings false. Being a generous host may simply have made the successful hunter more attractive.

As if all these incentives to food-sharing were somehow insufficient, *not* sharing food could also be perilous. Hunter-gatherer societies have always tended to kill, exile, or otherwise suppress their mute, inglorious Kozenys.

Inventing Agriculture

The step from intensive hunting and gathering to actual domestication of plants and animals didn't happen in a season or two. Like so many business decisions that look in retrospect like the product of genius, you would honestly have to say we stumbled into it. The most widely accepted explanation of how agriculture first appeared is that intensive methods had allowed some hunter-gatherers in the Levant to settle down and have larger families. When the climate suddenly worsened fifty or a hundred generations later, in the period known as the Younger Dryas, 11,000 years ago, the nouveaux intensives could no longer simply pick up and revert to the squirrel-skinning, berry-picking skills of some distant great-grandpappy. They needed a new way forward.

A few tribes figured out that survival lay in agriculture. Instead of merely gathering wild crops, they concentrated on plants with characteristics suitable for cultivation. Archaeological specimens of wild wheat seeds from the period tend to have a brittle rachis, the structure connecting them to the grass stem, and they shatter easily. Domesticated seeds, though, have a more durable rachis, suitable for transportation and storage back in the village, and they quickly become more abundant in the archaeological record. In wheat and some other species, the transition from wild to domesticated plants may have taken as little as 150 years.

From Brian Hayden's point of view, the trouble with the conventional explanation of the rise of agriculture is that it seems to rely on what biologists refer to as "group selection": The wise and self-sacrificing members of a tribe worked together to domesticate crops, with the evolutionary result that tribes of self-sacrificing sod-busters survived and gradually proliferated. The group selection scenario had considerable sentimental appeal in the 1950s and 1960s, with everyone tilling and hoeing for the common good, voices lifting in unison to sing some variant on "The Farmer and the Cowman Should Be Friends." But by the 1970s, the theory of group selection had fallen into disdain among biologists. Like the idea of having sex for the good of the group, it simply did not seem probable. Proponents of *individual* selection argued that the first question even the noblest among us asks in any situation is not What's in it for the group? but What's in it for me? In his 1976 book *The Selfish Gene*, Richard Dawkins said the question is actually, "What's in it for my genes?" Even the most altruistic behaviors in the natural world—the honeybee dying to defend her hive, the meerkat on his termite mound keeping lookout while his compatriots eat—come down, on close analysis, to what's likeliest to help the selfish individual's genes proliferate in future generations.

This brings us back to our Triple-A personalities, our mute, inglorious Kozenys, languishing among the tribal proletariat on a diet of grass seed and berries. Even before the appearance of agriculture, the new crowd of intensive hunter-gatherers displayed a powerful urge for privatization and they stayed in one place long enough to show it off. Status symbols crop up in the archaeological record for the Mediterranean Levant beginning about twelve thousand years ago: "decorated mortars, polished stone dishes and cups, stone figurines, decorated bone tools . . . and personal jewelry in the form of chaplets, diadems, frontlets, bonnets, bracelets, necklaces, and anklets. . . ." These glittery knickknacks testify to the ascendance of a self-aggrandizing elite. They weren't yet rich, but they did a fair job of acting rich. Hayden characterizes these Triple-A aggrandizers as eager "to maximize their power and influence by accumulating desirable foods, goods, and services"—that is, by hoarding. Like chimpanzees bringing meat to a feeding cluster, they also went out of their way to channel these goods through themselves as rewards to loyal followers. In particular, the Triple-A sorts began to ratchet up the cycle of competitive feasts.

Many archaeologists regard the proliferation of feasts around the start of the agricultural epoch as a result of improvements in food production. Hayden argues that they were also a cause. Giving a lavish feast and prevailing in "competitive battles" with rival feast-givers was important enough to drive the search for new and more impressive kinds of foods—not staples but status foods, not porridge in every pot or bread on every table but party foods.

Hayden points out that the first domesticated crops in many cultures around the world were actually intoxicants and delicacies. Some were even party utensils: In Japan, Mexico, and the eastern United States, one of the earliest domesticated crops was the bottle gourd, useful chiefly as a serving vessel at feasts. Elsewhere, it was the chili pepper—not a staple food but, as in some Maya cultures even today, a status symbol. Wheat may have been domesticated for bread, but some researchers say beer came first. And chickpeas? Think hummus and pita chips.

Comes the Revolution

At first glance, the idea that agriculture was a by-product of the frivolous quest for status flies in the face of reason. It seems much more logical that we took up farming to avoid the dire prospect of starvation. But it may be that we underestimate the deep primate yearning for status: People still literally starve themselves to death for the goal of looking good. Hayden suggests that we also underestimate the importance of feasts, not just for conferring and displaying status, but as an instrument for social manipulation. He regards the feasts of the period as a political necessity, a way to disarm potential adversaries as the Triple-A sorts began to accumulate personal wealth. What they were undertaking was the single biggest revolution in history. It was a break not just with our long history of more or less egalitarian tribal life but with the much older bias against hoarding behavior in primates. To get away with it, the Triple-A sorts used parties much as chimpanzees use feeding clusters, as a way to toss the rest of the tribe a few scraps. The feast, like the feeding cluster, was a promise: Stick with me, friend. We've got good times ahead.

Having taken control of tribal wealth, did the new elite deliver on this promise? Hayden writes that when he was studying tribal villages in the Maya Highlands of Mexico, he "was completely astonished . . . that the local elites provided essentially no help to other members of the community in times of crisis, but instead actually devised means of profiting from the misfortunes of others." Given free rein, he adds, Triple-A sorts "usually ruin the lives of others, erode society and culture, and degrade the environment." But free rein is not something humans readily give one another, and Hayden makes it clear that, one way or another, the promise of better times ahead actually got repaid a thousand times over: Once they had figured out how to domesticate crops for their prestige value, their party value, people realized that they could use these crops for practical purposes, too. Having brewed domesticated wheat into beer, they went on to knead it into bread. Thus we stumbled, perhaps literally, into the agricultural way of life.

The driving force of the Triple-A personality quickly latched onto the extraordinary surplus-making engine of agriculture, and then the revolution had come. In this Neolithic Revolution, as in the Industrial Revolution, almost every aspect of human life underwent change: Permanent settlements proliferated. Round huts and tentlike dwellings gave way to rectangular houses. Builders devised trial-and-error solutions to such rudimentary architectural challenges as the foundation, the freestanding wall, and the fireplace. Templelike cult buildings began to appear, and status and art objects abounded.

The rich, in Hayden's view, played a dual role in society. On the one hand, they were outlaws, pirates, pariahs, bending laws, exploiting neighbors, seizing every little advantage. On the other hand, they became the role models who produced, or served as patrons for, almost every great advance in the rise of civilization. The pattern of innovation that produced agriculture recurred endlessly over the millennia that followed: Triple-A sorts drove the development of new technologies for the purpose of displaying their prestige and rewarding followers. Then the sweaty mob imitated them, gradually figuring out how to put prestige technologies to practical everyday use. The earliest pottery from archaeological sites around the world, for instance, is typically highly decorated, designed for display at feasts. Workaday pots turn up only later. Likewise textiles, metalworking, leather shoes, indoor plumbing, and illuminated books appeared first as symbols of wealth, and only later evolved into necessities for ordinary people. One probably cannot attribute the rise of the World Wrestling Federation to the rich, but NASCAR should know that the first car race in America was sponsored by Alva Vanderbilt in 1899, on the lawn of her Newport mansion. It was an obstacle course of dummy policemen, nursemaids, and babies in carriages. Biographer Barbara Goldsmith writes that the driver who killed the fewest innocent bystanders won the race.

Inspiring envy and imitation was of course the reason the rich sought new technologies in the first place. Like being the chimp with the meat at the center of the feeding cluster, or the lady at lunch with the very latest gossip, it was a way to gain prestige and social control. But imitation also cheapens the symbols of power, Hayden writes, and "essentially forces successful aggrandizers to look for, or to develop, ever more costly prestige items." It was the start of a cycle of one-upmanship that remains the joy and the bane of life among the very rich unto the present day.

Viktory

At Peak House in Aspen, Viktor Kozeny's guests had mopped up their South Indian tea-smoked elk tenderloin with juniper glace and listened to Natalie Cole being seasonal. They'd savored Vietnamese coffee crème brulée for dessert, served with Château d'Yquem. By 3:00 A.M., everyone was headed home, but the party had in truth only begun. A few days later, Kozeny sat down with a local friend and pored through the snapshots his event photographer had taken. He was full of questions about who these people were and where they lived, ostensibly because he wanted to send them photos. Over the weeks that followed, he began to court a select group with lavish little lunches and dinners at Peak House. One day he had the place filled with a forest of orange trees, the next he had them swept away and replaced with palm trees because he was in an Arabian mood. He invited new friends down to his place in Lyford Cay and out on his 165-foot yacht *Contemplation*.

Kozeny of course had another investment scheme afoot, this time involving privatization vouchers in the former Soviet republic of Azerbaijan, on the Caspian Sea. He planned to use the vouchers to gain control of SOCAR, the national oil company, with its vast untapped reserves. Kozeny was too smooth to ask his new friends for money. Instead, he flew them around the world in his private jet, and wherever they went he played the generous host. At the best hotels and restaurants, it was always Viktor's party, and he swept up his guests in the excitement of his international wheeling and dealing. The SOCAR deal was a sure thing, Kozeny told them, because he was "in bed with" the president of Azerbaijan. Not a pretty place to be, under normal circumstances, but Kozeny claimed that SOCAR would yield up to a hundred times the original investment. Former U.S. Senator George Mitchell, a minor investor brought into the scheme by his Aspen friends, actually met with the Azeri president and confirmed Kozeny's prospects for conjugal bliss. The Triple-A personalities of Aspen lined up to invest. They probably ought to have been more circumspect. Just a year before Kozeny's fabulous Christmas party, *Fortune* had published a feature article describing the "painful lesson" previous investors had suffered at the hands of the man

Fortune dubbed the "pirate of Prague." The name Kozeny gave his Azeri venture, Oily Rock Group Ltd., might also have raised one or two well-groomed eyebrows.

In fact, what Kozeny raised was $450 million, much of it from Aspen residents and the Wall Street institutions to which his Aspen friends had introduced him. As *Fortune* later put it, "Kozeny may have been a pirate, but he was their pirate." His nonstop party promised good times ahead, and so his investors gave him free rein. The predictable trouble began as the SOCAR deal was due to be consummated. Azeri officials refused to accept Kozeny's vouchers, and the president, inexplicably getting out of bed, did nothing to change their minds. In August 2000, the vouchers expired, unredeemed, worthless. The ruined investors alleged in a lawsuit that Kozeny had actually bought the vouchers for forty cents apiece six months before his big party in Aspen. Then he sold them to Oily Rock for $25 each, skimming off more than $100 million in profits. Kozeny, back in the Bahamas, denied the allegations. He was, like his investors, simply an unfortunate victim of the wild mood swings of the emerging markets.

Anyway, he added, if he actually had $100 million in profits to spend, "there would be another party in Aspen." And given the continual pressure for one-upmanship among the rich, it would doubtless be a better party, at that.

4

Who's in Charge Here?

Dominance the Rough Way

I know you are the Prince of Wales and you know that you are the Prince of Wales, but the pig doesn't know that you are the Prince of Wales.

—Sir Pratap Singh *to his guest, the future King Edward VIII, who had made a dangerous blunder while out "sticking" wild pigs on horseback in India.*

In shallow pools along the shores of Lake Tanganyika lives an odd little fish, a type of cichlid. In this species, subordinate males are poor, harried creatures, sexually undeveloped, with drab, femalelike colors. But when a subordinate usurps a dominant individual, the newly anointed alpha male almost instantly turns bright yellow or blue. Over the next week or so, he experiences an eightfold increase in the brain cells that produce a key chemical governing sexual development. He acquires the cichlid equivalent of *cojones*. He also lays on muscle and sports a flashy red tip on his dorsal fin, like a pocket handkerchief in a double-breasted suit. In fact, the happy cichlid does everything possible to apprise rival males of his dominant status except put on epaulets and war medals. He does everything imaginable to attract females short of standing on his tail and fishlip-synching Barry White's "I'm Qualified to Satisfy You."

Dominance should be this easy in humans. But it is the bane of the modern hotel concierge that the guy in the T-shirt and tattered blue jeans may, in fact, be a billionaire. And it is the soul-rending dilemma of the *maitre d'*: Does one furtively seek secondary dominance signals before fawning—a slight bow to disguise the glance down in search of the $2,000 Berluti shoes? A two-handed welcome, to see if the incoming wrist perhaps bears a Patek Philippe watch? Or must one actually be nice to *everyone*? Philip Anschutz, whose fortune is estimated at $9.6 billion, offers little help: He wears a dime store Timex. Likewise King Abdullah II of Jordan sometimes likes to ghost around his country in rags and New Balance sneakers.

This is what makes dominance such a tantalizing concept in humans. It is almost invisible in our daily lives and yet also present everywhere. (Try showing up late for an appointment with Anschutz. Try stepping on Abdullah's sneakers.) In every situation, people size one another up as quickly and ruthlessly as grade-schoolers choosing sides in a pickup basketball game. Every time two people meet, some scientists say, the question of dominance or submission gets answered in the way one person holds eye contact and the other glances away, or in the way one unconsciously shifts vocal tone and body language to match the other.

The Meaning of Dominance

Dominance is by no means the exclusive prerogative of the rich. Humans are an extraordinarily variable species, and it is entirely possible, as Mohandas Gandhi demonstrated, for a poor person to humble the mighty despite being short, slight, badly dressed, physically unattractive, and philosophically opposed to the use of force. But by and large, wealth, the control of resources, is a good determinant of who's in charge around here. An ordinary individual who doesn't like the way the game is being played typically has little choice but to take his ball and go home. A rich person, on the other hand, can buy the ball park and shoot the referee. A professional polo player once told me about a stint in the employ of an Asian potentate, who was watching a chukker from the sidelines, accompanied by his pet cheetah. When a referee made several flagrantly impartial rulings, the cheetah inexplicably came unleashed, bolted onto the field, knocked the referee from his horse, and stood on his chest in an advisory capacity .

There are in fact ways—some of them subtler than a hungry cheetah at a subordinate's throat—by which the rich routinely let everyone around them know precisely who is in charge here. There are rules by which they gain and keep the upper hand in any situation. Before we open this rule book, it will help to understand something about social hierarchies in the natural world—and also about the many caveats the idea of dominance seems to evoke.

Pecking Orders

Dominance is an astonishingly new idea in biology, first put forward early in the twentieth century by an obscure Norwegian researcher studying the behavior of chickens. Thorleif Schjelderup-Ebbe coined the term "pecking order" and defined the idea of the dominance hierarchy. For his efforts, he was thoroughly crushed by the Scandinavian biological hierarchy and never gained academic employment. But his ideas caught on. In the 1930s, other researchers coined the term "alpha male" to describe the leader of a wolf pack. Biologists began to see dominance hierarchies almost everywhere.

Unfortunately—and this is the first caveat—they have never come remotely close to agreeing on just what dominance means. Different researchers looking at the same group of animals may identify the dominant individual by any of four common definitions: It's the one who can beat up everybody else but doesn't necessarily need to (the Warren Buffett 800-pound gorilla style of dominance). Or it's the one who displays the most aggression (the school of competition exemplified by Larry Ellison of Oracle, who once quoted Genghis Khan: "It is not sufficient that I succeed; everyone else must fail"). Or it's the one to whom other members of the group pay the most attention (the Richard Branson "Look at me in a wedding dress" paradigm). Or it's the one who gets the first pass at resources like food, sex, or a nice place to sleep (King Fahd of Saudi Arabia has bedrooms reserved for his pleasure in the palaces and yachts of

Saudi princes around the world, the equivalent of several dozen $2,500-a-night hotel suites, on call every night, all year long, year after year, though he will almost certainly never visit them.)

The obvious problem with these four definitions is that the same animal can easily turn up as the alpha in one study and the beta in another. Researchers who regard control of resources as the defining factor, for instance, tend to believe the dominant individual in a group is *least likely* to display aggression. Everybody else is too scared to challenge his status, except at great intervals. Aggression is more typical, they say, of middle-rank individuals jockeying for position in the hope of an eventual bid for the top.

Figuring out who's in charge is thus more complex than the casual language of alpha males and pecking orders might lead us to think. The way dominance gets expressed varies from species to species, from individual to individual within a species, and even, for a given individual, from day to day. Irwin Bernstein, a University of Georgia psychologist, argues there is little evidence that dominance has *predictable* effects on behavior in *any* primate species. A dominant rhesus macaque with other things on his mind (or a full belly) may let a subordinate take away his meal. A female bored by her alpha male may slip away to frolic with a hot young beta, a behavior Bernstein's staff refers to as "going on safari."

The one thing most biologists agree about is that dominance isn't a personality trait. It's not some posture or presumption that makes a person the boss in any situation. Rather, dominance is a relationship between two individuals. A boot camp corporal may be dominant when he's barking orders at a recruit, but not when he's saluting a master sergeant. Bill Gates may lope like an alpha through the hallways at Microsoft, "but for all I know," says Bernstein, "his chauffeur browbeats him." Unlike animals, humans don't generally spend the bulk of their lives in a single herd or pack. We move routinely from one hierarchy to another, from imperious chairman at a Fortune 500 company to chastised parent of a discipline problem at Greenwich Country Day School, from old money in Cleveland to new potatoes (rather small) in Palm Beach.

In fact, the more I talked to researchers and rich people alike, the more it seemed that they were in a kind of collective denial about the whole idea of dominance. The rich claimed never to exercise power in any sense more forceful than "responsibility" or "leadership," or, a favorite euphemism, "governance," much as they eschewed any real interest in money. Larry Ellison maintained that he had quoted the line about how "everyone else must fail" only in the spirit of stern disapproval for such unseemly zeal. (This was presumably during a brief period of moral reform when he was not urging employees to "kill" the competition or hiring spies to root through the garbage of his corporate enemies.) And the biologists? Something inherently disturbing about the idea of dominance seemed to make their legs go wobbly. One biologist talked about "latency to emission of terminating responses" when what he meant was "how long it takes till the loser backs down." Wading through the literature, I sometimes got the feeling that dominance was as abstract as black hole theory, until suddenly something reminded me that it can be as raw as the memory of having been pushed around in high school. So maybe I shouldn't have been surprised to discover that no one has ever attempted a systematic study of dominance in adult humans, much less rich adult humans. It's a taboo.

The Power Hum

Yet dominance is almost as basic to primate life as breathing, and perhaps as subconscious. One intriguing study suggested that we unwittingly declare our dominance or submission every time we open our mouths. Researchers from Kent State University taped twenty-five interviews on the *Larry King Live* talk show, paying particular attention to frequencies below 500 herz. In the past, most researchers disregarded these low-frequency tones as meaningless noise, a low, nonverbal humming on which the spoken word rides. But as they toted up their results, sociologists Stanford Gregory and Stephen Webster noticed that in every conversation the low-frequency tones of the two speakers quickly converged.

This convergence seemed to be essential for a productive conversation: The speakers literally needed to be on the same wavelength. But it wasn't simply a matter of two people finding some happy middle ground. In talking, as in walking, one person set the pace: King's low-frequency tones shifted to the level of his guest when he was interviewing someone with high status like Ross Perot or George Bush. On the other hand, lower-status guests tended to defer to him, "but with less gusto," the authors noted. The most deferential guest was former Vice President Dan Quayle.

Gregory and Webster, who have since duplicated their results in other contexts (to weed out the Larry King suck-up factor), theorize that our vocal undertones provide a means by which we routinely and unconsciously manage "dominance-deference relations." This nonverbal form of communicating status, says Gregory, may be why one person overhearing another on the phone can tell by tonal qualities alone whether the speaker is talking to a boss or a friend. The low humming beneath our words seems to be, as an anthropologist once put it, "an elaborate code that is written nowhere, known by none, and understood by all."

An Elaborate Code

So how do we read the rest of the unwritten code of social dominance? An alpha wolf establishes his authority by biting a rival on the neck and pinning him to the ground. Rich humans are seldom quite that direct. On the Mexican border one time, I met a beautiful, fast-talking, finger-snapping young woman in spangled earrings and a short skirt. She let me know unabashedly and at every turn the power of her family's money. "I mean, we don't own *everything*," she said. "But we own things that you guys need. That humans need. We own the gas company. We own the industrial park. We own the print shop. We have a construction company. Nobody has anything against us," she said, and then she explained: "*We'll cut off your gas.*"

The better class of rich people tend to frown on such crass expressions of power. Well, they frown on them and yearn for them at the same time. People expect Oracle and Microsoft to work out their differences like grownups, in the courtroom, with antitrust lawyers as intermediaries. But how delicious it is when the rivalry surfaces in a personal context. A couple of summers ago, Oracle's stock was soaring and Larry Ellison briefly passed Microsoft cofounder Paul Allen to become the world's second-richest person. Ellison was celebrating aboard his 243-foot yacht off Capri. Then he spotted a 200-foot yacht heading out on a twilight cruise to the village of Positano. It was Paul Allen's *Meduse,* the sort of thing rich people make a point of knowing. When they land the Gulfstream V at Aspen airport, they check out the tail number on the Gulfstream V next door to see how the pecking order stands. So Ellison ordered his captain to crank his yacht's three engines to full speed. He overtook Allen's yacht at forty miles an hour, throwing up a huge wake that sent Allen and his guests staggering. "It was an adolescent prank," Ellison told *The Washington Post* afterward. "I highly recommend it." It was of course also an expression of social dominance. (Foul weather warning: Allen has since moved back up to number three on the *Forbes* 400, while Ellison has fallen behind to number four, and both men now go down to the sea in much bigger ships.)

As in other species, human dominance may derive from a reputation for ferocity, but it can depend just as powerfully on a reputation for philanthropy. For our alpha males and females, the cues to social dominance can include physical bearing, steadiness of gaze, a style of dress, a family name, an aura of wealth, a big house, a network of influential friends. Odd as it may sound, dominance may sometimes be simply a matter of knowing how to be nice. It may, in fact, be all of the above. Wealth allows an individual to accumulate the tools of social dominance in almost unlimited variety and depth, to be deployed as needed.

Deployment generally follows two basic patterns, one aggressive, the other gentle; one based on wielding a big club, the other on belonging to the right club; one forceful, the other seductive. It is of course much easier to live with the latter than the former. For rich people of this sort, dominance can become an art form in which the goal is to get one's way with the least possible effort. They may remind subordinates of their status through little more than the quiet language of fine clothing and expensive wristwatches. Or they may deliberately avoid even these little hints. (The person who does not wear the right watch or the four-button suit wins points for being modest. He also plays a game of "gotcha" with strangers who must figure out on their own just where he fits in the social hierarchy. When King Abdullah of Jordan dresses down, it isn't so he can savor the experience of being a normal person, leading a normal life. It's so he can catch lazy bureaucrats and sack them.)

Unfortunately, we must leave this subtler brand of dominance for a later chapter because it rides, however elegantly, on the back of the harsher, more aggressive style of dominance. That is, it depends ultimately on the implicit threat of force, however remote. "Monaco," a violinist in the national orchestra there once told me, "has a first class symphony, a first class ballet, a first class opera. Everything here is a jewel," and then he added, as if the one thing somehow flowed from the other (and it does), "We have the most phenomenal police department. There is no crime." He might also have added that there is no dissent. Anyone who offends the Grimaldi Family—Prince Rainier, Prince Albert, or the Princesses Caroline and Stephanie—risks being escorted politely and permanently to the border. Every velvet glove must have its iron hand. So in the remainder of this chapter, we'll focus on the kind of dominance that typically creates wealth, the raw, first-generation, force-of-nature style of dominance, as practiced mostly by the happy brigands of the business world.

Monomaniacs

Back in 1980, when Ted Turner was attempting to establish his Cable News Network as the first twenty-four-hour national television news service, I asked him to predict the half-dozen or so most powerful people of the coming decade. He named one. "Obviously I couldn't very well name myself," he began, tentatively. "But if I have my way . . ." The "if" was also a little provisional, though he was of course talking to a stranger with a tape recorder, at a time when most Americans had not even heard of cable television, much less installed it in their homes. Then Turner hit his stride: "I have never met—and I'm trying to be modest, OK?—I have never met anyone yet who has studied the entire future as carefully or with as much research or in as much depth as I have. I got a better handle on it than anybody." This was pure "mouth of the South." And of course he said it outright: "I think I'll have more impact than any other single individual." I rolled my eyes and got out of his way, which is pretty much what everybody else did. CNN is now a world power, and, with a personal net worth fluctuating between $3 and $9 billion, so is Turner.

Did he prevail because his vision of the future was clairvoyant, and is remarkable foresight or intelligence a necessary characteristic of the dominant personality? I don't think so. Visions are cheap, even very good visions like Turner's. (Indeed, one Turner ally at CNN has claimed that the original idea for a twenty-four-hour cable news network came from *Time* executive Gerald Levin, who let the opportunity slide.) On the other hand, almost all successful alpha personalities display a single-minded determination to impose their vision on the world, an irrational belief in unreasonable goals, bordering at times on lunacy. For instance, when an obscure Cleveland businessman named John D. Rockefeller set out to create his Standard Oil cartel in 1870, the petroleum industry was a sprawling, chaotic mess characterized by wild financial speculation. Rockefeller, by contrast, was in many ways a colorless figure with a "hyper-trophied craving for order," in the words of biographer Ron Chernow. But Rockefeller also had "a wide streak of megalomania." Right from the start he confided, "The Standard Oil Company will some day refine all the oil and make all the barrels."

The common tendency is to treat this kind of unstoppable determi-nation among business leaders as a personality defect, the product of a warped childhood. One biologist recently went so far as to suggest that in modern human society the alpha personality is "deviant." Classifying alphas as defectives or deviants is in truth little more than a strategy sub-ordinates use to help themselves cope with the demands of the people who run their lives. It is a kind of hapless shrug: "He's crazy, what can I do?" People with this perspective sometimes act as if they have explained Rockefeller's drive simply by pointing out that his father was a scoundrel, or as if Ted Turner became a media giant only because his father was a drunk who sometimes whipped him with a coat hanger. But children of scoundrels and drunkards can just as easily end up being scoundrels and drunkards themselves, and children from healthy families can become tycoons. Bill Gates had a warm relationship with his parents and now employs his father to run his philanthropies. Richard Branson writes, "I cannot remember a moment in my life when I have not felt the love of my family." They may sometimes be odd, intensely success-oriented fam-ilies. Donald Trump's father, for instance, had the habit of chanting, at every conceivable opportunity, "You are a killer . . . You are a king." But this kind of background predicts almost nothing. It can turn one kid into The Donald and another into a drunk. We don't know what blend of nature and nurture produces dominant individuals.

What we do know from studying other primate groups is that dominance behavior is entirely natural; there is nothing defective or deviant about it. Unfortunately, the origins of dominance are just as vague in the animal world. Writing about the alpha chimps she has known, Jane Goodall notes that all but one shared "an intensely strong motivation to dominate their fellows." (The exception got the job on account of being too big for anybody else to intimidate.) They became dominant, apparently, because they wanted to dominate—really badly. They wanted it badly enough to persist in spite of repeated failure. Goodall describes one alpha making the same dominance display nine times in fifteen minutes before finally shaking a rival out of a tree.

What's more interesting is that both human and chimp dominants seem to want it with a reckless, all-or-nothing passion that defies the basic sense of self-preservation. Goodall describes one newly anointed alpha named Mike being stalked by five angry males. Mike flees into the trees, then suddenly turns on the lot of them and stands his ground. The shocked posse runs yelping for safety. It's an instance "of the importance of psychological factors in chimpanzee dominance interactions," according to Goodall. "It implies also that the lone male who dares to face such opposition is either stupid (cannot imagine the possible consequences) or has rather a large share of boldness—a quality that perhaps comes close to courage."

Goodall leans toward boldness over stupidity, but it isn't necessarily an either/or proposition. A blindness to consequences characterizes many dominant personalities. They don't see obstacles. They may not bother or even be capable of seeing other peoples' point of view. They tend to think, as Aristotle Onassis once put it, "The rules are, there are no rules." This blindness is what makes them, with equal ease, outlaws and heroes. In truth, the "maniacal" or "megalomaniacal" zeal with which a Ted Turner or John D. Rockefeller or Mike the Chimp pursues his vision is often what really impresses the rest of us and causes us to step aside (or even follow behind). In *The New New Thing*, his book about Internet mogul Jim Clark, Michael Lewis put it perfectly: "He was the guy who always won the game of chicken because his opponents suspected he might actually enjoy a head-on collision."

A Reputation for Upright Behavior

If dominance can be as quiet as the lift of an eyebrow, the turn of a vowel, the curt flicking of an index finger, it can also be as loud as Henry T. Nicholas, the forty-three-year-old cofounder of Broadcom Corporation. Nicholas is six-foot-six, broad-shouldered, with a goatee and fierce, impatient eyes set deep under a furrowed brow. As if this were not sufficient to make adversaries cower, he likes to be photographed from below, with arms folded and the winglike door of his black Lamborghini Diablo Roadster open behind him. (Read: Big, bad, and in a hurry.) This is one of the cruder facts of social dominance: Bigger people tend to get their way. They make more money, as Nicholas, who cofounded his company in a spare bedroom and sold $500 million of stock at the height of the Internet boom, clearly knows. They are disproportionately represented among the ranks of those who run the world. In Tahiti, the tribal chiefs, or *ariki,* were so much larger and more corpulent than their subordinates that visiting Europeans assumed they were a different race (if not quite a different subspecies). Likewise, in a 1980 study, when the average American male was just five-feet-nine-inches tall, more than half the chief executive officers in Fortune 500 companies were six feet or taller.

There are, to be sure, exceptions to the close connection between height and social dominance. Silvio Berlusconi has made himself one of the richest men in the world, and twice been elected prime minister of Italy, despite being just five-foot-six. But even Berlusconi understands the value of at least appearing larger than life. Before he sits down at a conference table, an aid puts a cushion on his seat to jack him up to the height of his colleagues. For group photos, Berlusconi surreptitiously goes *en pointe* when the cameras click. Dominant humans typically stand straighter and move more expansively, much as the alpha male in a wolf pack walks with head and tail erect. The smaller ones may also snarl more. Hollywood agent Michael Ovitz once greeted an employee's six-foot-three-inch boyfriend this way: "You're too tall for this business. You need to get into another business or cut your legs off, because you're a threat to me."

Size matters. Even Henry Nicholas works to enhance the effect. He belongs to that underrated category, the big man with a Napoleon complex. The guy who always has to prove he's smarter and stronger than everybody else. When Nicholas donated $1.28 million to the crew team at the University of California at Irvine, for instance, he showed up in suit-and-tie for the presentation, just a middle-aged businessman in a monogrammed shirt. He chatted with the young rowers, lean, powerfully built kids with strapping shoulders, and then, because this is the sort of dumb, spur-of-the-moment thing Nicholas likes to do, he challenged them to a pull-up match. With television cameras rolling, he offered an extra $1,000 for every pull-up "you can do more than me . . ." and, in case they didn't understand just how easy this was going to be, he added, "Remember, I'm an old man." (He was in fact about to turn 40.) The team captain took the challenge, and only then did Nicholas do a quick Superman-strip to reveal a very cut, muscular torso. The kid managed 13 pull-ups. Nicholas ran off 27, reversing his grip to eke out the final few meaningless points. It was, a school official admitted afterwards, "kind of a setup." Like dominant animals in the natural world, Nicholas has a knack for entering contests he knows he can win.

The Winner Effect

Winning little contests is almost a prerequisite for winning big ones, which may be why the self-made rich sometimes compete so fiercely for even the most trivial distinctions. It has to do at least partly with testosterone, which occurs in men at the minuscule rate of a hundred-thousandth of a gram per liter of blood (and about one-seventh that level in women). Testosterone has become notorious as the supposed cause of aggression, a sort of "raging bull" hormone, but studies have repeatedly demonstrated that testosterone levels in a group of men actually predict nothing about who will display aggression. In one study of grade-schoolers, boys with a history of physical aggression actually had lower testosterone levels at age thirteen; they were also failing in school and unpopular with peers. Elevated testosterone levels weren't associated with aggression at all, but with social success—that is, with dominance.

Testosterone levels in men rise in anticipation of almost anything that can be interpreted as a dominance contest, whether it's a rugby game or a chess match. For the losers in these contests, according to Allan Mazur of Syracuse University, loss of status typically results in relatively depressed testosterone levels. After the 1994 World Cup soccer tournament, for example, fans of the losing Italian team experienced decreased testosterone levels, and fans of the winning Brazilian side enjoyed a testosterone spike. Mazur theorizes that increased testosterone encourages an individual to put on the signs of dominance—erect posture, sauntering gait, direct eye contact. This may make for greater success in subsequent dominance encounters: "Success begets a high (testosterone) response which begets more dominant behavior which begets more success." Biologists refer to this as "the winner effect."

Hence the vital importance of a supposedly impromptu pull-up contest for Henry Nicholas. The "winner effect" from these little contests may help set him up for success in much larger contests. He needs all the help he can get because his larger goals are huge. He recently told *Worth* that he aims to run six miles in less than thirty-six minutes, bench-press 350 pounds, and lead Broadcom to $10 billion in gross annual revenue. These three goals, *Worth* pointed out, constitute "a fiendish paradox" as "each one . . . directly opposes the other two. You can't go on distance runs without burning the muscle mass you need for bench-pressing; you can't work 90-hour-plus weeks and still have enough sleep for your muscles to rebuild." To lesser morals, the goals Henry Nicholas has set for himself look, in a word, maniacal. We tend to get out of his way.

Dhoom-dham

If bigger and stronger are not enough, Nicholas, like many other dominant animals, is also louder. In the animal world, making deep, loud noises is one way males tell the world just how big they really are. Deep sounds require large body spaces to serve as resonators. So the roaring of howler monkeys and the belching of bullfrogs are a reliable way to scare off rivals and attract the attention of discriminating females in search of Big Daddy. Biologists who study frogs refer to this as "deep croak," a concept that also resonates in the world of the rich and powerful. Great fortune-builders are also often great screamers, and the diatribe has been a favorite tool of David Geffen, Ted Turner, and Intel's Andy Grove, among others. Henry Nicholas, who frequently works eighteen-hour days, is known for scheduling staff meetings at 11:00 P.M., often pushed back to 3:00 A.M., at which he rages, growls, and curses at his weary employees.

For a dominant animal, the noise one can make with one's own body cavities is merely a starting point. In the natural world, orangutans do not merely bellow but also push down huge dead trees to announce their progress through the forest. It's a kind of deep croak behavior by proxy. Indian princes used to do the same sort of thing, with criers and heralds anticipating their passage through the streets by chanting their titles and honors. The Hindi term for this clamor was *dhoom-dham*, and its function was to remind not just the people but also the prince of his regal status. Henry Nicholas does much the same thing with a Harley Davidson and the Lamborghini, in which he plays the heavy-metal band Metallica loud enough to be audible over the growling of the engine. He once staged a battle-of-the-bands at his 15,000-square-foot house in Laguna Hills, with a $2,000 prize to make it more interesting, and for his fortieth birthday party he hired a band called Orgy. The dhoom-dham was good, and the neighbors had to call the police.

In primatological circles, Mike the Chimp is famous for having achieved dhoom-dham, and social dominance, by banging together two kerosene cans. He was a young, unimportant male on the big day, and, according to Jane Goodall, he started by carefully studying a group of adult males, his social superiors, who were grooming nearby. Then, his hair sleek and his manner relaxed, Mike walked over and picked up two empty cans outside Goodall's tent. He carried them back to his previous place and sat down to stare at the other males. They ignored him. Mike began to rock from side to side, his hair slightly erect. No one paid any attention. He rocked more vigorously and his hair became fully erect. Suddenly, he charged at his superiors, uttering pant-hoots and clanging the empty cans together with both hands. The other males panicked and fled. It took Mike another four months to get all the males in his troop to acknowledge him as alpha, but he did it without ever risking a physical fight. The combination of pant-hoots and clanging cans did the job.

Like Mike, Henry Nicholas cultivates the deep croak image as a dominance strategy. He will sometimes crush an employee in front of a reporter, who gets to feel that gleeful, horrified sensation of having stumbled in on a nasty domestic spat. But it isn't by accident. Nicholas and his victims sometimes stage and script these public executions in advance, so he can appear even louder and more brutal in the press. It's a way to convince employees, investors, and, above all, rival companies that he and Broadcom will go any distance to win. He puts on what his wife (gamely) calls "his game face." Ideally, he also wins without a fight. Demoralized rivals never even enter the competition.

The Plus Face

How else does the unwritten code of social dominance speak to us? The body language of dominant individuals is at times almost a self-fulfilling prophecy. In one study of small children, dominant individuals typically entered a confrontation with eyebrows raised and chin up, making full eye contact with their opponents. This so-called plus face helped produce a win 66 percent of the time; the minus face, with eyes down and chin lowered, produced a win less than 10 percent of the time. The plus face actually seemed to stop subordinate children in their tracks. They reacted to a minus face in 0.64 seconds; it took them almost twice as long to muster any response to a plus face. Studies like this unfortunately focus on children because adults are so much more secretive about dominance and more likely to alter their behavior when being observed.

But evidence suggests the body language stays the same. Indeed, even before it meant "kingly," the word "rich" meant "to lead in a straight line." For the rich, going directly for what they want, without hesitation, becomes a habit, natural and without self-consciousness. Or as one woman from a wealthy family put it, "A friend of mine in college once said that I walk through a door like I never had one shut in my face, and it's true. I'm basically very comfortable in the world." The college friend no doubt resented it on some level. Yet not displaying this kind of expansiveness and self-assurance can clearly be hazardous to a leader's health. In October 2000, for instance, Jeffrey Henley, the chief financial officer of Oracle Corporation, walked into a meeting with stock analysts and reported that company sales were on target. But Oracle is known for bluster and bravado, and Henley displayed what CNBC's *Squawk Box* characterized as "cautious body language." It gave a new meaning to the idea of the minus face: Oracle's stock promptly dropped 12 percent, and another 12 percent the next morning. If, as one behaviorist has written, the whole notion of dominance behavior is "an invention" existing only "in the mind and notebook of the human observer," our minds and notebooks nonetheless count, in this case to the tune of $118.4 billion in market capitalization lost at least partly on account of insufficient swagger.

Eye contact also counts. One way to assess who's dominant in a conversation, according to Glenn Weisfeld, a psychologist at Wayne State University, is to consider the ratio of "look-speak over look-listen." Dominant individuals generally make eye contact when they're speaking, but look away when subordinates are speaking to them. If this sounds disrespectful, says Weisfeld, that's because "disrespectful is precisely what dominant individuals do."

Staring at another person communicates that the individual doing the staring wants something and is not merely determined to get it, but entirely confident in his ability to do so. He doesn't need to glance over his shoulder for reassurance. He doesn't fret about being attacked from behind. In the animal world, markings around the eyes exaggerate the direction and intensity of the stare. For instance, the lines that run under the cheetah's eyes and focus down its nose do not, as cat-lover's like to aver, make it "the cat that cries." What those lines really say is, I am the cat who is going to eat you for dinner.

Dominant humans use their eyebrows to produce roughly the same effect, and their prominent noses to point the stare unmistakably at their prey. Donald Trump likes to leave his eyebrows untrimmed because he thinks they make him more intimidating in negotiations. In case this might be too subtle, an admiring gun maker once designed a revolver for Trump with the title of his book *The Art of the Deal* inscribed around the rim of the barrel, so one could read it only when the gun was pointed between one's eyes. The design also included a silver bullet with Trump's face carved on the tip. (With remarkable restraint, Trump never actually had the gun made.)

The photographer Edward Steichen likewise testified to the power of a dominant individual's gaze, after making his famous portrait of the American financier J. P. Morgan. Staring into Morgan's eyes, he said, was "like looking into the light of an oncoming express train." Over the years, people routinely praised Steichen's insight in depicting Morgan not merely glowering but with a dagger in one hand. This was, according to Steichen, "their own fanciful interpretation of Morgan's hand firmly grasping the arm of the chair." But it is testimony to Morgan's predatory reputation and to the intimidating effect of his stare. Morgan himself tried to buy the first print for $5,000 from the collector Alfred Steiglitz, and later had Steichen make him several prints. Even now, ninety years after Morgan's death, the portrait is a way to reproduce the moment of rabbit-panic in the hearts of potential rivals he never had the pleasure of meeting, as it were, eye-to-eye.

The Honorable Predatory Impulse

A muted urge to act like big, fierce animals often seems to rumble at the subwoofer level through the lives of the rich: A purr on the cusp of becoming a growl. The idea in the animal world is to develop a reputation for ferocity, and you will seldom have to defend it. It works the same way for the rich. Having committed one or two notoriously ruthless acts, they may achieve dominance thereafter without doing anything, without in fact even knowing that something might need doing. Their reputation alone defeats the competition.

A few years ago, for instance, Microsoft scheduled a gala rollout for a new version of its Windows operating system at the Moscone Convention Center in San Francisco. The San Francisco area happens to be home to Sun Microsystems and its chairman Scott McNealy, a billionaire in whom Bill Gates induces an acute sense of relative deprivation. McNealy has described Gates as "the most dangerous and powerful industrialist of our age." So McNealy's marketing executives had a bright idea. Microsoft products are notoriously riddled with bugs. Why not remind people of this embarrassing record by hiring a fleet of yellow pest-control trucks to circle the convention center during Microsoft's big party? The term for this sort of behavior in the business world is "mooning the giant," defined as a foolish act intended to antagonize a large, threatening rival. The Sun Microsystems executives had twenty trucks lined up around the corner from the convention center and ready to roll. Then the local office manager for Western Exterminator suddenly realized that he was being hired not by Microsoft, but by some rival whose name he mistook for "Microsun." On the other hand, he knew perfectly well who Bill Gates was, and he'd heard about his penchant for extermination on the grand scale. Gates, he said, could buy his company and close it. The pest-control boys promptly rounded up their trucks and headed for the hills. McNealy must have wept.

Grooming a reputation for ferocity is thus a serious business. Balzac wrote that behind every great fortune there is an undiscovered crime. (This at least is how we remember the phrase in our timorous hearts. What Balzac actually wrote in his novel *Pere Goriot* was more nuanced: "The secret of a great success for which you are at a loss to account is a crime that has never been found out, because it was properly executed.") Balzac assumed that the beneficiaries of the crime would want to keep it secret. But many families actually celebrate the seminal crime, at least after the passing of a two- or three-generation sort of statute of limitations. The sacred image of the family's former rapacity becomes a tool for intimidating rivals down the generations.

In Monaco, for instance, everything begins and ends with the Grimaldi family's palace, home to the mild-mannered princes Rainier and Albert. It is a natural clifftop fortress. The Grimaldis, originally from Genoa, first got inside when a thirteenth-century forebear posed as a Franciscan monk begging for alms. The guards opened the gates to give comfort, whereupon the ancestral Grimaldi and his henchmen rushed in and slaughtered them. It was the sort of treacherous episode one might expect the family to suppress, especially as the Grimaldis now pride themselves on their charitable endeavors. But in fact the family shield features the motto "Deo Juvante," or "With the Help of God," over an image of Franciscans bearing swords.

References to what Thorstein Veblen called "the honorable predatory impulse" are a standard tool of social dominance. In the past the usual practice among the rich was to invite guests to join in an actual hunt as a display of predatory power and an act of blood bonding. Fox hunting in particular became emblematic of wealth. The rich got to romp across the countryside in archaic aristocratic clothing and commune with the spirits of dead kings. It was an opportunity to display prowess on horseback and genuine skill at working with hounds. Fox hunting is of course no longer politically correct, and hunters nowadays generally play down the blood-shed. The serious ones say they hunt to "watch the hounds work." The less serious ones say they just like a wild gallop across open country. Hardly anyone admits to the raw appeal of the predatory impulse.

But in the hunts I've witnessed, it's the possibility of death that causes pockets of momentousness to open in the course of an ordinary winter morning. A horse may dig in at the last moment at a stone wall. Eyes glint, nostrils flare, hooves skid in the wet earth, and the horse crashes and somersaults over. A rider who ends up at the bottom of the heap on the other side is said to have been "buried" by his horse. An apt phrase. Foxes die, too, rather more often. One of them crosses a wetland patch with the pack yelping frantically ten yards behind. The gap closes. Suddenly the fox evaporates in a cloudburst of blood. For novices in on their first kill, a swipe across the cheek with the fox's bloody tail is the tribal ritual of initiation: The mark of the predator.

Even if the great-grandchildren have gone over to the antihunt side, they often hang onto reminders that their ancestors at least were big, fierce animals. A hunt scene painted by Gericault, say, will certainly appeal on the strength of artistry and sheer spectacle. It also reminds vis-itors that they have entered the territory of a killer. At one house I visited, the main stairway featured a Frans Snyders painting of hounds and a lion literally red in tooth and claw, opposite a Snyders canvas of a bear fling-ing aside a hound, and a canvas of a lion sinking its teeth into the flanks of one man while another man lay fallen with his belly clawed open. Amid this carnage hung serene portraits of formidable ancestors.

Paintings by George Stubbs are generally more subdued, but they can convey the same subliminal message. For instance, over the couch in the smoking room of the private apartments at Blenheim Palace, where the duke and his family relax, there's a Stubbs painting of a recumbent tiger. The tiger was given to the family by Clive of India in the eighteenth century, and when I visited Blenheim two-and-a-half centuries later, a member of the staff was still ritually recounting its diet: The tiger ate twenty-four pounds of meat every two or three days, plus a cow's head once a week. "It helped to discourage poachers," she joked.

The honorable predatory impulse also turns up in less obvious contexts, as a kind of continual subtext in the lives of the rich and powerful. On the windowsill in Jerry Della Femina's office in mid-Manhattan, for instance, someone has taken a set of silver block letters spelling his last name and turned it into an anagram: INFLAME DALE. This is entirely appropriate. Della Femina, who owns his own advertising agency and two popular New York restaurants, is a provocateur. He was once arrested for displaying pumpkins outside a gourmet store he owns in East Hampton, a violation of the community's straitlaced design codes, and he prides himself on being an ethnic Brooklyn boy lighting a fire under the WASP snobbery of the Hamptons Old Guard. But when I pointed out how appropriate the idea of inflaming the dale was, Della Femina acted as if he'd never seen the thing before. Some tools of social dominance are perhaps better kept subliminal.

Fierce Furniture

A taste for fierce imagery also turns up in the furniture of the rich—for instance, in the ball-and-claw foot of a Chippendale chair or the crossed-spears on a Colonial American bull's-eye mirror. These devices may seem like little more than attractive design motifs, but they are rooted in the ancient aristocratic custom of glorifying aggression, through heroic sagas, paintings and military memoirs, family coats of arms, and almost any other conceivable means. The ethologist Irenäus Eibl-Eibesfeldt points out that it was once fashionable for men in Austria, Germany, and Switzerland to have themselves ornamentally scarred with a duelling sword, like Waika Indians of the Upper Orinoco, who used to shave their heads to display scars acquired in club fights.

No one ever used aggressive imagery more enthusiastically than Tipu Sultan, an eighteenth-century Muslim ruler of Mysore in southwestern

India who waged a bloody war against the British invaders. Tipu made the tiger his cult animal and built a reputation as "the Tiger of Mysore." Chained tigers guarded his palace on the fortified island of Seringapatam. His legend-makers asserted that when two of his ministers displeased him, he fed them to the tigers.

While these tales have their undeniable charm, what's more pertinent to the lives of modern-day rich people is the way Tipu's real predatory behavior became ritualized into a decorative motif. Within the palace, Tipu's throne was a sort of dais raised up on tiger legs with huge clawed pugs (much like the ones still seen on some dining room sideboards). The full figure of a tiger stood underneath in the middle and the tiger's golden head projected forward toward approaching supplicants with its mouth open and fangs bared. Eight more tiger heads topped the corners of the dais. To drive home the predatory message, Tipu wore tiger stripes himself and used them on the upholstery of his throne and in his chambers.

The crowning touch was a sculpture in the music room, three-quarters life-size, of a tiger sinking its teeth into the throat of a European soldier. The tiger's body concealed a mechanical device, which produced tiger snarls and the agonized screaming of the victim. In addition, the tiger contained a fourteen-pipe organ with a keyboard to provide "either supplementary discords or a triumphal air to follow." Scholars believe the human figure may have represented either the British colonel whom Tipu decisively defeated in the bloody 1780 battle of Pollilur or the son of the commanding general, who was actually mauled to death by a tiger some years later. Either way, the memory of Pollilur, kept alive by this tiger and by a series of wall paintings at Seringapatam, helped convince Tipu of his own ferocity. According to a biographer, "He thus acquired a self-confidence—not to say arrogance—in his dealings with the English which only deserted him in the closing scenes of his life." Indeed, it's doubtful that it deserted him even then, as he died fiercely slashing his sword at the grenadiers overrunning his palace in the British assault of 1799.

The value of Tipu's decorative motifs as symbols of dominance has persisted well beyond his death. The golden tiger head from his throne now bares its fangs in Windsor Castle. The tiger eating the soldier is now a trophy of the Victoria and Albert Museum. The eighth Baron Harris of Seringapatam and Mysore, descendant of the general who defeated Tipu, still carries the image of Tipu's tiger (its heart pierced with an arrow) on his coat of arms.

The predatory impulse also remains an essential tool in high fashion to this day. Like Tipu, the rich wear tiger stripes, leopard spots, snakeskin, crocodile, and other animal patterns not just because these materials are scarce and precious, but because they evoke a predator alert response in everyone the wearer meets. Leopards, for instance, have been preying on humans and on our prehuman ancestors for at least two million years, and this unhappy experience may well have shaped the evolution of our aesthetic sensibilities. Some primates have evolved a visual system that is highly sensitive to the color yellow, apparently for easier detection of leopards. A piece of spotted fur no larger than a football still produces alarm even in bonnet macaques which have not lived with leopards in more than a hundred years. It is in their genes, and apparently in ours, too. Richard Coss at the University of California at Davis has argued that just such a genetic heritage may be the reason kings, tribal leaders, dictators like the late Mobutu Sese Seko in Zaire, and fashionable women all tend to favor leopard-print clothing. Leopards still make us a little gaga. They command our attention and our respect. We cannot help it.

Verbal Phallocarps

We have at this point only begun to explore some of the cruder tools of social dominance deployed in pursuit of wealth. Questions of size, upright posture, and prominent pointing suggest that we conclude for now with a brief aside on the topic of penis display behavior. This may or may not be one of the great canards of the social dominance debate. In his 1969 book *The Human Zoo*, Desmond Morris suggested that even highly civilized men use ritualized penis display, or penis display by proxy, as a means of social dominance: "The tough, dominant male . . . who chews on his fat cigar and thrusts it into his companion's face, is fundamentally performing the same Status Sex display as the little squirrel monkey that spreads its legs and thrusts its erect penis into the face of a subordinate."

It is, regrettably, difficult to avoid the impression that rich men, particularly the self-made ones, sometimes treat ritualized forms of penis display as a synecdoche for social dominance. They build phallic symbols, as when Donald Trump casts the seventy-story shadow of yet another Trump Tower on the glitterati of midtown Manhattan. (Trump says that if you measure the newly erected Trump World Tower his way, taking into account the high ceilings, it's really as big as a ninety-story building.) They set out, like Internet billionaire Jim Clark, to create a yacht with "the biggest mast ever built," at 189 feet. (And then some other rich guy goes and tops it, with a 220-foot "stick" currently under construction.) The rich also frequently deploy verbal phallicisms, as when producer Jeffrey Katzenberg, preparing to release the film *Dick Tracy*, wrote to a rival, "You won't believe how big my Dick is." Indeed, it sometimes seems that the rich are actually trying to *become* phallic symbols. They want to be Big Swinging Dicks, most famously on the bond-trading floor at Salomon Brothers in the 1980s: "A new employee, once he reached the trading floor, was handed a pair of telephones . . ." Michael Lewis wrote in *Liars' Poker*, an account of his time at Salomon. "If he could make millions of dollars come out of those phones, he became that most revered of all species: a Big Swinging Dick." In those politically incorrect times, managing directors used the phrase to compliment successful traders, as in "Hey, you Big Swinging Dick, way to be." And Lewis added, "To this day, the phrase brings to my mind the image of an elephant's trunk swaying from side to side. Swish. Swash. Nothing in the jungle got in the way of a Big Swinging Dick."

The elephant trunk analogy is oddly misplaced, and it begins to suggest some of the hazards in the phallic mythology of the rich: The elephant has a penis, too, and it also goes swish, swash. So much so that when a bull elephant is in the periodic state of heightened hormonal activity called musth, he is in considerable danger of stepping on it. It is otherwise an impressive sight. "An elephant's penis weighs 45 kilos," the woman who discovered musth in African elephants once told me, "and so do I." At which point there was a pause in the conversation.

It gave me a moment to reflect that the Big Swinging Dick is a highly relative concept. Human males in fact have a disproportionately large penis compared to other primates. By contrast, silverback gorillas weighing 450 pounds have penises measuring less than two inches when fully erect. But before we break out the Macanudo cigars for a round of congratulatory narcissism (and a moment of silence in honor of the intrepid biologist who defended human masculinity, and imperiled his own, by actually getting close enough to measure a gorilla's erection), we should also keep in mind that one of the most sexually endowed creatures in the world is the flea. Its penis is not only one-third the length of its body but is also equipped with claspers, feather dusters, and french ticklers. Modesty perhaps becomes us.

In any case, why should any rich man particularly feel the need to demonstrate that his penis, or proxy penis, is bigger than the next guy's? Fleas aside, aren't we all relatively well endowed? Part of the answer is that we aren't competing against gorillas, but against other human males. During the course of our evolution as a bipedal species, the penis was a highly visible means for attracting the attention of females and also for impressing other males. One female anthropologist has even argued, a little improbably, that our species began to walk upright in the first place as a better way for males to show off their penises. (In many other primate species, males temporarily become bipedal when they want to show off. But surely there are better reasons to walk on two legs?)

Our primordial great-grandmothers, the Madonnas and Peggy Guggenheims of their day, may well have chosen males based in part on penis size, much as our randy great-grandfathers focused on breasts. This wasn't necessarily because a bigger penis (or breast) made for better sex. Sexual selection is seldom that straightforward. Women may actually have focused on penis size the way elk does focus on big antlers in stags, as an indicator of overall quality in a male. Female sexual selection often favors male traits, like big antlers, bright colors, long tail feathers, or megalomaniacal ambition that are valuable chiefly for intimidating other males.

Would an organ as laughable as the penis intimidate anyone? Vervet monkeys in East Africa apparently think so. They have a red penis prominently displayed against a brilliant blue scrotum. When a vervet troop is feeding, some males will sit facing away from the group with their legs spread. If an unknown member of the species approaches, these guards get an erection and put on a threatening face. Some humans do the same sort of thing. A few years ago, *National Geographic* ran a photograph of tribesmen in Irian Jaya dressed in what the embarrassed caption-writer called "modesty sheaths." The phallocarps, or penis sheaths, in question were in fact up to two feet long, brightly painted and ornamented with tufts of fur and feathers. This wildly immodest display was apparently intended to impress and intimidate rivals.

If such displays are a natural behavior for humans and other primates, what happens when we put on our pants and pass laws against indecent exposure? Zoo animals denied the opportunity to do what comes naturally often pace restlessly or engage in other forms of so-called displacement behavior. And humans? One possibility is to wear tight pants and perhaps supplement what comes naturally, like the heavy-metal musician with the cucumber down his pants in the mockumentary *Spinal Tap*, or for that matter, like King Henry VIII, who wore a suit of armor in which the codpiece was prominently upright. Another possibility is to build skyscrapers.

This still begs the question of why a larger penis, or penis proxy, would translate into social dominance. In vervet monkeys, the erection display may be what Nancy Etcoff, a psychologist at Harvard Medical School, calls "an action manqué—a ritualized threat to mount." Mounting, in this context, isn't a sexual overture at all, but what Desmond Morris refers to as status sex, an act from which the sexual meaning has been stripped away, leaving only the dominance. The opposite behavior, submission, also takes a ritualized form: Among monkeys, putting oneself in a position to be mounted is a commonplace gesture of submission. Youngsters present themselves rump-first to their mothers, and adults of both sexes do it to dominant males. It's no more than a polite way of saying, You're the boss.

Etcoff's ritualized threat to mount may be only one side of the dominance-by-penis-display story. Though one does not like to think about it, Melissa Gerald, a primatologist at the University of Puerto Rico, points out that penis competition works the other way around, too. Just as subordinate humans suffer from the minus face in dominance confrontations, they may also be afflicted with the minus penis. In males who have been frightened or defeated, the testicles typically retract and the penis dwindles. So while our naked ape ancestors doubtless wished, like bond traders in the 1980s, to display the Big Swinging Dick, they may have worried just as fervently about not showing the tiny shriveled one.

In fact, the more closely one looks at the whole idea of penis display the more it comes to look rather unfortunately like a can of worms. It should be obvious first of all that not every vaguely phallic object, like a cane or a dangling necktie, actually constitutes a ritualized form of penis display. Old men use canes because they need them. Bruce Springsteen wears string ties because he thinks they look good, not to cast aspersions on his own manhood. Developers sometimes build taller skyscrapers not to show up the guy next door but because it's the best way to maximize square-footage and personal net worth, a slightly more refined way of measuring up. Moreover, even in the more unmistakable cases of phallic symbolism, like the swagger sticks, scepters, batons, maces, and truncheons ceremonially displayed by the rich and aristocratic, it is wishful thinking to imagine that every penis display necessarily translates into dominance. It depends on the context.

In his book *Old Money*, Nelson Aldrich, Jr., matter-of-factly described the tendency of some rich men "to pee in open fireplaces, potted palms, ladies' reticules, any convenient place so long as it be in full public view." Aldrich deemed the practice common enough to merit a name, "urination syndrome," and listed newspaper owners Ned McLean of *The Washington Post* and James Gordon Bennett of *The New York Herald* among the "victims." Two instances of this odd behavior suggest just how important context can be.

The eccentric tenth Duke of Marlborough was once chatting in his usual mumbling style with a visitor to Blenheim Palace, the lovely and elegant Lee Bouvier Radziwell. Without warning, he paused a moment to urinate in the nearest fireplace, then finished his sentence with a characteristic "What?"

A more poignant and revealing instance of urination syndrome involved the flamboyant New York millionaire Evan Frankel, best known for producing the musical *Brigadoon* on Broadway. Late in life, Frankel fell permanently in love with a beautiful young woman named Elena Prohaska, and for several years they lived together at Frankel's East Hampton estate. Later, Prohaska became engaged to Burt Glinn, a Magnum photographer, and she brought him to meet Frankel for his blessing. The meeting went badly. After a drink, Frankel, who was by then in his eighties, took Glinn off on a tour of his extraordinary estate, which included winding trails, fanciful statuary, a "no tops, no bottoms" swimming pool, and a peacock shed. (The frequent appearance of peacocks on the lawns of the rich serves as a subtle advisory to anyone who has somehow missed the point that these estates are about extravagant male display behavior. Like clock towers, the peacocks have the advantage of periodically calling attention to themselves with their music, which resembles a chorus of large, healthy babies being thrown one by one off a cliff.) "She could have had all of this," Frankel told Glinn, a half-hour into the tour. Then, instead of anointing the younger man as a worthy successor, he unzipped his fly and made a show of peeing on the side of the peacock house.

In both cases, urination syndrome was a ritualized way of degrading the viewer. For the Duke of Marlborough, it was evidently an impersonal act, no more than a class habit of treating the lower orders as if they are not quite there. Frankel's act, on the other hand, reeked of animus and petulant submission, like a baseball manager kicking dirt on the umpire after being thrown out of a game. It certainly wasn't an act of dominance.

To put this aspect of penis display in perspective, let's finish, as we began, with squirrel monkeys. It turns out that Desmond Morris was wrong when he suggested that tough men and squirrel monkeys thrust their erect penis-equivalents into the faces of subordinates as an act of dominance. The squirrel monkey, at least, does no such thing. This tiny New World primate is the sort of animal zoo-goers invariably call "adorable," with a small, expressive face, big eyes, large tufted ears, and a skittish disposition. (A bit like Amazon.com's Jeff Bezos, come to think of it.) Relative to its body size, it has the longest penis of any primate. But Morris' interpretation of what it does with this endowment was based on the behavior of squirrel monkeys in captivity. Given the conditions in zoos at that time, this was a bit like interpreting human behavior based on the prison population at Devil's Island. Since then, Sue Boinski, a University of Florida researcher, has spent years observing different squirrel monkey species in their native habitat in Costa Rica and Suriname. And, yes, a male will sometimes thrust his erect penis in another squirrel monkey's face. But the clear context is not dominance. It's submission. "It's a sign given by subordinates to dominants," says Boinski. "It's to say, 'I acknowledge that you're bigger and meaner than me, so we don't need to fight about it, particularly since I would almost certainly lose.' Like a dog lying on his back with his belly exposed to attack: 'O.K., bite it off.' "

This is something to think about (though only in the spirit of sympathy) next time a tough, dominant male chews on a cigar and thrusts it in your face.

5

Take This Gift, Dammit!

Dominance the Nice Way

The brain is unreplaceable, [but if necessary] I'd ask for one that came from a smart dog.

—William H. Gates, *age ten*

[I was like] a rabbit that's small and fast. All my big competitors were like a pack of wolves, and they were all chasing me, but I was fast enough to be out in front of them.

—Ted Turner, *age sixty-three*

In the dusty pink hills of an Israeli desert, a bird called the Arabian babbler worked his way along the small branches of a dead tree, hammering here and there with his beak. This particular bird's name (at least in the minds of the biologists who studied him) was Tasha-Sham and he was a brownish, forward-leaning bird, about the size of a mockingbird, with a long tail and a sleek head. Unflamboyant. Even drab. Unless you happened to be a birder, you probably wouldn't travel ten minutes out of your way to see a babbler in the field. Tasha-Sham hammered a little harder, eagerly stripping away bark until he plucked out an amber prize, the fat, succulent, terrified larva of a beetle. Then he did a remarkable thing. Instead of gobbling it down, Tasha-Sham flitted up onto a tree where Pusht, the beta male in his group, had been doing sentinel duty.

Pusht saw what was coming and slipped away. But Tasha-Sham, who was the alpha, followed him to the ground with the unmistakable bearing of the magnanimous fellow bent on committing acts of goodwill. He held up the prize morsel until Pusht dutifully begged like a nestling, mouth agape, wings quivering with feigned enthusiasm. Then Tasha-Sham crammed the larva down Pusht's throat.

It was, on the face of it, a puzzling piece of self-sacrifice. But the logic of his actions did not trouble Tasha-sham for a moment. He drew himself up to his full height and announced his gift by lifting his beak in a special trill, almost a purr, like a socialite posing for an event photographer at the Breast Cancer Awareness barbecue. Ah, sweet charity.

Arabian babblers seem to understand the value of doing nice things for one's fellow creatures. Nor are Arabian babblers the only philanthropists in the natural world. In fact, altruism is remarkably common among animals. Even some social amoebae do it. Free-living amoebae, at that, which do not normally get together even for sex. They reproduce by cloning. But when times are bad, they clump together and about 20 percent of them go on to amoeba glory by turning themselves into a lifeless stalk to support the others. So it shouldn't be too surprising that rich people practice altruism, too.

The puzzle is why. In the narrowest Darwinian terms, acts of charity make no sense. A trait should survive and proliferate only if it benefits the individuals who display it. Giving away food or other resources appears to do just the opposite. Altruism is risky behavior. It represents an apparent reduction in Darwinian fitness, a loss in an individual's ability to survive and reproduce. Let's put Tasha-Sham's little act of philanthropy in perspective: You are stuck in the desert. You don't know where your next meal will come from. You make a living by banging your head fifty or a hundred times against a tree trunk until, miraculously, you come up with Beluga caviar. Then you give it away? And to a subordinate? Shouldn't animals like Tasha-Sham quickly starve to death? If this is what altruism's all about, shouldn't it long ago have dwindled away and disappeared from the gene pool?

The rich face no imminent danger of starvation, but there are at least three special reasons philanthropy might seem to make no sense even for them. First, for people who tend to define themselves in financial terms, loss of net worth can translate into loss of status. In a widely noted 1996 diatribe against his fellow billionaires, Ted Turner blamed the *Forbes* 400 list for plutocratic stinginess. "That list is destroying our country!" he complained to *The New York Times*. "These new superrich won't loosen up their wads because they're afraid they'll reduce their net worth and go down on the list. That's their Super Bowl." Turner was of course wrong to blame it all on the *Forbes* 400. The philanthropic stinginess of the rich has venerable roots; rich people have often paid lip service at best to charity, while giving heart and soul to the preservation of capital. The Duke of Marlborough's family, for instance, used to practice charity by heaping their leftovers in tins, with everything from soup to trifle "in horrible jumble." The Duchess and her children would then accompany this benefaction to be ceremonially slopped out at the homes of poor families in the Blenheim Palace neighborhood. When the heiress Consuelo Vanderbilt married into the family, she introduced a radical note of American magnanimity—separating different kinds of leftovers in different tins, "to the surprise and delight of the recipients," as she boasted afterward. Another aristocratic family shared the spirit of Christmas by opening the windows of their townhouse and pelting food down on poor people waiting in the street. Sharing their muttonchops doubtless gave them a nice, warm feeling (especially if they hit their targets) and posed no threat to net worth.

But let's move on to the second reason sharing real wealth can be a horrifying idea: Stinginess is often part of the mindset that makes people rich in the first place. One night at a charity event in Palm Beach, I sat next to a woman in a hot pink Escada evening gown with black polka dots, her breasts served up like ripe fruit, her hair swept round in an ice cream whirl. She told me that she drove a new Jaguar; that she had places in Palm Beach, New York, and the Hamptons; and that her gown had cost $4,000. A little later, after two or three glasses of wine, she confessed that her fortune had come from a rubber-stamp business founded by her immigrant father. He was so frugal that he used to patch the torn seat of his office chair with Scotch tape rather than waste money replacing it. He kept the heat in his building so low that when his daughter did office work for him she had to wear a hat and gloves with the fingertips cut off (style by Dickens, not Escada). His money was no more than an abstraction, and the idea never entered his head that he had enough of it to heat his own building, much less enough to give away to heat other peoples'. Everything went into the company.

The third reason charity makes no sense for the rich is that hanging onto every penny can seem like the key to building a family dynasty. A recent ad by one brokerage firm depicted a founding father, nouveau riche but tasteful, with this thought in mind: "I will have great-great-grandchildren who come from 'old money.'" And it went on to say, "They aren't even born yet. And I can't imagine what their names will be. But, in addition to my own retirement, it's for these future generations that I treat my investments with such care today." The urge to build a dynasty, to be the founder of a great family, is entirely natural. Monkeys do it, too. Vervet grandmothers in high-ranking families orchestrate advantages for little Tiffany Vervet and Percy Vervet III. When play gets rough, the alpha mom rushes in to make sure her child prevails. Other monkeys learn to treat the offspring of high-ranking families more carefully. Social dominance thus tends to get inherited. Even creating a device like the generation-skipping trust—to ensure the orderly passage of wealth down the generations and to protect the principal from divorce lawyers, tax collectors, and other potential interlopers—is a natural behavior, though somewhat beyond the scope of the average vervet. The founding father can't go on living forever, but if he hoards sufficiently and plans well, he can ensure that his children and even his great-grandchildren survive and reproduce in comfort. It's what Darwinians call kin selection. By taking care of his descendants and close relatives, the rich man improves his own genetic fitness, propagating his genes and perpetuating his name through his lineage.

What seems at first glance to be distinctly unnatural is choosing to help strangers at the expense of one's own progeny.

Ted Turner, for instance, has five children and loves them all, in his way. (Turner, who has belatedly become an outspoken proponent of population control, once told a reporter, "I can't shoot 'em now that they're here.") He also possesses an ego of the sort that might naturally lend itself to dynasty-building. But in September 1997, more or less on the spur of the moment, Turner announced that he was planning to give away $1 billion, which at that point represented a third of his net worth, to strangers. Specifically to the United Nations, to address issues like population control and epidemic disease. If his children had done the math in the sober light of the next morning, when they got the happy news, it might have gone something like this: Subtract a third of the family fortune, take away half of what's left in estate taxes, divide by five, and in theory each of them was now likely to inherit no more than $200 million, down from $300 million the day before. Some of them may even have looked ahead and calculated that, if they had three kids apiece, the next crop of Turners would get less than $70 million each. Better than a sharp stick in the eye, to be sure, but a long way from the *Forbes* 400. At this rate, to take the worst possible case, some not-so-distant Turner could end up as a wage-slave for the spawn of archfiend Rupert Murdoch, who did not give away a third of his net worth.

So why would someone as shrewd as Ted Turner take the chance? Turner himself readily admitted that self-interest played a part. "I have learned," he said, "the more good I do, the more money has come in." There is nothing shameful about this, nor does it call into question the sincerity of his belief in the causes he supports. Any realistic point of view, and certainly the Darwinian one, recognizes that all philanthropy has a strong element of self-interest. In Los Angeles, where people tend to like their ambition naked, a businessman named Robert H. Lorsch happily admitted that he gets back $1.01 to $2 for every dollar he spends on charity. He doesn't necessarily give with this thought in mind. It just works out that way. Philanthropy positions the donor favorably in the public mind (Lorsch has his name over the entrance to the local science center) and brings him into contact with the right people. In one case, Lorsch's donations to cancer research led to his becoming an early investor in a pharmaceutical start-up. Lorsch figures that his million-dollar investment has already increased severalfold, with an eventual thirtyfold payoff when the company, CancerVax, goes public.

Altruism would not be so popular, among either Arabian babblers or billionaires, if it did not suit self-interest. The interesting question is how. Acts of charity obviously serve the rich as a form of advertising. Having your name carved in granite on a museum or university is a way to announce to posterity that you are a person of the highest possible taste. Getty made his fortune in oil, Guggenheim and Frick in steel, Hirshhorn in uranium, and Mellon in aluminum, but we've largely forgotten their skill at ripping commodities from the Earth; instead, we admire them for having created some of America's greatest art collections. It amounts, at times, to money laundering. The name Stanford University now suggests intellect and integrity, not the political chicanery on which Leland Stanford built his fortune. A. Alfred Taubman got rich as a developer of suburban shopping malls. Then he went on to own Sotheby's, the art auction house, where he was recently convicted of colluding with competitor Christie's to rig commission levels and gouge fellow plutocrats. But he is far likelier to be remembered as the benefactor of the Taubman College of Architecture and Urban Planning, the A. Alfred Taubman Health Care Center, the Taubman Medical Library, the Taubman Center for Public Policy, and the Taubman Center for State and Local Government. There is so far no plan to endow a Taubman wing at Allenwood federal prison.

Philanthropic advertising also clearly functions on a more personal level. Giving makes the donor more attractive. It is the rich person's equivalent of a peacock's tail. Buying a $500 ticket, or a $5,000 table, at a charity dinner announces both that you can afford the donation and that you are generous enough to undertake it. Rich and with a kind heart— the perfect mate. This was almost certainly the message my friend in the $4,000 Escada evening gown was hoping to communicate. Her act of philanthropy, like her dress, was a mating display, and in case there was any doubt, she told me three times in the first five minutes of our conversation (twice with her hand on my arm, once with her hand on my back, and always with her breasts beneath my nose) that she was looking for a husband. The stuff about the stingy father who ate up her youth came out only after I told her I was already married.

Even anonymous gifts can be a form of advertising. A rich man doesn't necessarily need everyone in the world to know about his benefactions. It may be enough for his wife to know, or for his mistress to come across a thank you note nonchalantly tossed on a side table. "The greatest pleasure I know," Charles Lamb once wrote, "is to do a good action by stealth and to have it found out by accident."

But is it plausible, as one writer has recently suggested, that Ted Turner's monumental gift was merely a courtship display? In his book *The Mating Mind: How Sexual Choice Shaped the Evolution of Human Nature*, Geoffrey Miller writes: "We could ask, from a Darwinian viewpoint, why men should bother acquiring more resources if they just end up giving them away. One clue emerged in a Larry King interview. Turner revealed that when he told his wife of his intended gift, she broke down in tears of joy, crying, 'I'm so proud to be married to you. I never felt better in my life.' At least in this case, charity inspired sexual adoration." Or as R. G. Ingersoll wrote in his *Tribute to Roscoe Conkling*: "We rise by raising others—and he who stoops above the fallen, stands erect."

But I don't buy it. The idea that any man would undertake a billion-dollar mating display for a woman to whom he is already married, even if the woman happens to be Jane Fonda, seems, oh, unlikely. And ineffective, too, since Turner and Fonda subsequently divorced. At the risk of sounding terribly middle-class, I have a hard time getting my mind around the idea of a billion-dollar mating display for anybody. (Personally, I lean more toward the style of King Victor Emanuel II of Italy, who used to give his mistresses one year's growth of his big toenail, though I think he went a little far in having it polished by a jeweler, framed in gold, and studded with diamonds.)

So how else to explain Turner's gift? Consider the context: Why do men and women go on acquiring resources, as Miller suggests, well beyond any conceivable need? The answer is of course for the sense of control, accomplishment, and success in social and financial competition. Or as Richard Branson has put it: "Every second I'm *fightingfightingfighting* and *competingcompetingcompeting*. Can't waste a minute." This isn't a frame of mind that easily changes overnight. The rich thus often end up giving away resources for the same reasons they acquired them in the first place. It's about *winningwinningwinning*, about doing just a little better than your nearest natural rival, the guy who gives you that uneasy sense of relative deprivation.

Charity, in other words, is often about social dominance.

Smart charities understand this and thrive by inducing a sort of arms race among potential donors. Thus Harvard University did not merely tell Albert J. Weatherhead III that he was "Harvard's third largest living donor," but also shamelessly added, "and you won't rest until you are No. 1." Not *we* won't rest, *you* won't rest. Weatherhead, a big man in Cleveland, where he owns a molded plastics company, recently gushed to *The New York Times,* "I just love that quote." So does Harvard. Weatherhead has so far forked over $50 million. Nor has it been a one-sided transaction. Three endowed professorships at Harvard now bear the Weatherhead name, as does the Weatherhead Center for International Affairs, which focuses on "the major issues and pressing challenges confronting the world," thus ensuring that Weatherhead will be a big name for centuries to come, and well beyond Shaker Heights.

To understand this style of dominance behavior, let's go back for a moment to babblers and the puzzle of how altruism could have survived the harsh regimen of evolution. The theory of kin selection suggests that altruism has persisted because altruists do nice things mainly for close relatives, thus improving their own genetic fitness. But with babblers, as with humans, the beneficiaries of charity are often unrelated to the donors. Alternatively, the theory of reciprocal altruism suggests that altruism has survived because babblers who do nice things will get nice things done for them in turn, tit for tat. But the very idea makes babblers murmur, "*Quelle horreur.*"

Their altruism flows only one way. Tasha-Sham gives food to Pusht; but for Pusht to give to Tasha-Sham would be the worst sort of social affront. Despite their drab appearance, an almost obsessive concern with status colors every aspect of babbler life, and they are acutely aware that altruism is a favorite way of showing it. One day I watched a female babbler with a prize piece of food race off to find the other female in her group. These two birds, half-sisters, had a tangled relationship, each under the impression that she was the boss. When the younger bird offered to feed her, the older sister simply froze, as if to say, This can't possibly be happening. It was a charitable standoff. For babblers, giving is an honor, and getting is a disgrace. The babblers will eagerly accept food from the biologists who study them, because biologists have no place in the babbler social hierarchy. *But not from her.*

A popular story about the nineteenth-century financier Baron James de Rothschild suggests that it works the same way for humans. Delacroix had supposedly asked Rothschild to pose as a beggar because he had "exactly the right, hungry expression." The verisimilitude was such that when Rothschild dressed in rags for the part, an unwitting young friend of the artist gave him his spare change, a single franc. Rothschild subsequently had one of his liveried servants repay the handout with his version of spare change: A sum of 10,000 francs. The story is apparently apocryphal. An alternative version turns up in an Oscar Wilde short story, "The Model Millionaire," apparently based on Alfred Rothschild, one of James's English cousins. But the sense of muted indignation about this violation in the one-way character of charity rings true: You don't give to me, I give to you.

For babblers, altruism isn't the only means of establishing one's status. Both male and female babblers sometimes also practice aggressive dominance, chasing off rivals and ostracizing or even killing them. But aggression is dangerous and expensive; a bird can spend its whole day waging turf wars. It can of course also die in combat. Once the hierarchy of a group is in place, altruism is a far less expensive way to maintain the social order. A gift every now and then serves as a reminder that the alpha is in charge, and simultaneously gives the betas a little prize to console them in their status: You may be the underdog, but at least you've got a full belly. Dominant babblers don't give away every succulent larva they come across; they aren't dumb enough to starve themselves. They give just enough to keep the peace and maintain their status. "Sometimes to amuse myself, I give a beggar a guinea," Nathan Rothschild once confided to a friend. "He thinks I have made a mistake, and for fear that I should find him out, off he runs as hard as he can. I advise you to give a beggar a guinea sometimes; it is very amusing." Among other things, an act of charity intimidates potential rivals, especially if they know they cannot match it. It helps confirm the donor's dominant status, and more status typically means more mating opportunities. So charity does not, after all, reduce their genetic fitness; it enhances it. Thus altruism survives the harsh winnowing of evolution and proliferates.

The Charm Offensive

Rich people, like babblers, have different means of getting their way. They can do it by aggression and risk incurring the wrath of the lower orders, or they can do it by what some psychologists call "prosocial dominance," that is, the charm offensive. Both styles of dominance are constantly in play, but even social scientists tend to recognize dominance only when some alpha swings a big stick, and not when he or she speaks softly. In one recent study, for instance, researchers found that "socially dominant" men were much more likely to die prematurely. And good riddance: The study defined dominance as Type A behavior—monopolizing conversation, interrupting others, and competing too hard for attention. But we seem to recognize in our daily lives that these domineering traits are just as likely to indicate *lack* of dominance. When Soviet Premier Nikita Khrushchev beat the podium with his shoe and threatened to "bury" America, he frightened people—and undermined his own credibility. John F. Kennedy, by contrast, smiled a lot, displayed grace under pressure (backed up, to be sure, by a discreet display of intercontinental ballistic missiles)—and prevailed. Some social scientists now suggest that humans practice dominance most effectively not by bullying people but by doing favors, by philanthropy, by sharing attention, by building alliances, and by deploying the gentle weaponry of compromise and persuasion. Evolutionary psychologists, who are not otherwise known for their sunny view of human nature, suggest that we sometimes gain power by being, for want of a better word, *nice*.

This brand of dominance represents a radical transformation from the snarling tyranny of an alpha male in a pack of wolves or the top mouse lording it over his trembling colony. But it's a transformation that's increasingly evident as we move up the evolutionary ladder from simpler species to more complex and intellectual ones, particularly among primates. It's also a transformation some scientists see being recapitulated every day in nursery schools around the world. (We must resort once again to studies in children because there is no comparable work on adult humans. But preschoolers are a better model for the behavior of the rich than one might think.)

Patricia Hawley, a psychologist at Yale University and Southern Connecticut State University, writes about how dominance evolves as children mature. In one study, she paired dominant and subordinate children and gave each pair a game to play. Hawley chose the games, a little diabolically, to allow one player to do the fun stuff and the other the scutwork. Not surprisingly, the dominant children hogged the fun stuff.

In the early years, toddlers lack the verbal and social skills to get their way by subtle means, so they use force. As in a wolf pack, other toddlers pay more attention to aggressive individuals and prefer them as social partners. But it's different in older children. In one of Hawley's videotapes, a dominant five-year-old girl and a subordinate boy are playing with a toy fishing rod. The girl takes the first turn and then, as the boy tries his luck, she leans in and offers advice. "Should I help you?" she asks after a moment, gently taking back the fishing gear. Then she says, "O.K., we've caught your fish. Now it's my turn again." The two smile and remain friendly. But the girl controls the fishing 80 percent of the time.

For Hawley, this is what dominance is all about as we grow older (and, I would add, as we grow richer). The bully who grabs the fishing rod and the child who gets it by grace and good manners aren't all that different. Both aim to control a resource. They've just figured out different ways to do it. They may even be the same child, a few years apart. Sometime between first and third grades, says Hawley, children who continue to be bullies lose status. The ones who remain dominant figure out that they need to acknowledge the thoughts and feelings of their playmates.

This isn't, of course, the same as giving in to the thoughts and feelings of their playmates. Hawley describes the girl's strategy at the fishing pond as "a sophisticated way to dupe your partner under the guise of helping." Dominant individuals learn to use prosocial techniques like bargaining, compromise, cooperation, and appeals to friendship as ways to maintain goodwill while still monopolizing the resources. They are *manipulatively* nice.

Traditional animal behaviorists would say this isn't dominance at all. It's cooperation. Hawley replies that cooperation as practiced by our fellow primates is often a form of competition: Chimpanzees, for instance, form alliances and use them to gain status; bonobo monkeys trade sexual favors as a way to win friends and influence fellow primates. Humans may practice dominance by force, by simply taking what we want, but we also do it by giving away what we already have. We are the most intelligent life form on Earth, and also the only species to have evolved the transforming power of language. It would be surprising if we *hadn't* come to practice dominance through kind words and other forms of verbal manipulation. It would be odd if we didn't use all the accomplishments in our power—wealth, strength, audacity, education, even humor—to produce deference in others.

If all else fails, we can of course still snarl and thump our adversaries on the skull. But among humans, aggression is the crudest instrument of social dominance, and with the least predictable consequences. Microsoft's reputation as a corporate bully has helped make Bill Gates the richest man in the world, but it also provoked the U.S. government to sue Microsoft for antitrust violations. When Microsoft seemed to be recovering from its legal troubles, even *Fortune* resorted to a horror-movie coverline: "The Beast Is Back."

Consider, by contrast, the consequences of a little known but far more "prosocial" strategy Gates employed more than 20 years ago, when he and Paul Allen were just starting Microsoft. Allen, the technological wizard, had already dropped out of college and gone to work developing the company's software on salary from their first big client. It was another six months before Gates also took the plunge. Both men contributed equally to the founding of the new company, but when they formalized their partnership, Gates insisted on a 60–40 split in his favor, according to Gary Rivlin in *The Plot to Get Bill Gates,* on the grounds that Allen had been getting a salary for six months, and he hadn't. Also, Allen had dropped out of Washington State University. Gates would be dropping out of Harvard.

Gates's strategy was devoid of overt aggression, relying entirely on persuasion and the illusion that he was being, if not nice, at least fair. Moreover, Paul Allen certainly prospered from the partnership. He has always ranked just a step or two behind Gates himself among the richest people in the world. But by the beginning of 1999, according to Rivlin, the unequal partnership had cost Paul Allen $15 billion in lost profits.

You could call this cooperation, or you could call it the single most effective act of dominance behavior in the history of animal life on this planet. And not a fang was bared in anger.

Ted and Bill and Rupert

Let's get back to philanthropy, and the question of why someone as shrewd as Ted Turner would take a billion-dollar flyer on the United Nations. Context clearly mattered. Much as he had predicted in 1980,

Turner had in fact made himself one of the most powerful businessmen in the country, achieving his apotheosis with the rise of CNN during the Gulf War of 1991. But at least one other business figure to emerge in the 1980s could rival Turner's promised impact. The personal computer was as unlikely a medium in 1980 as cable television, but Bill Gates had arguably studied the future at least as carefully as Ted Turner. The 1980s also saw Rupert Murdoch expand the American end of his media empire with the purchase of Twentieth Century-Fox and Metromedia. Both men were natural rivals for Turner. By his own account, Turner had nurtured his megalomania with more than one hundred viewings of the movie *Citizen Kane*. He was, according to his family, both fascinated and horrified by the story of media giant Charles Foster Kane, closely modeled on newspaperman William Randolph Hearst, who ends up owning everything but dies alone. In 1996, Turner sold his media empire to Time Warner. He became vice chairman at the new company and its single largest individual stockholder, but he was clearly no longer The Man. With an ego as vast as Turner's, something had to give.

His relationship with Rupert Murdoch quickly turned to loathing. The two men were both single-digit billionaires, both sixtyish, both in media, both with large families at work in their empires (Murdoch had four kids then, but he has since had a fifth, matching Turner's five). Moreover, Murdoch had just begun to tread on Turner's CNN turf by creating his own cable news network. He wanted Time Warner to carry Fox News on its New York City cable system. From Turner's perspective, it was as if Murdoch wanted to have a baby just like Ted's, and, oh by the way, wondered if Ted's wife could carry it for him. When Turner and Time Warner refused, Murdoch used his political clout to herd the mayor of New York City, the governor, and the state attorney general into the dispute on his behalf, ostensibly in the cause of free speech.

With characteristic belligerence, Turner publicly likened Murdoch to the Führer. Murdoch replied with a muted threat to "get even." The rivalry was at its nastiest during the 1996 World Series, which featured Turner's Atlanta Braves and was televised by Murdoch's Fox Network. During the Series, Fox avoided showing Turner's image except when his team was scored against, a shun strategy common among kindergartners and other bands of unruly primates.

Turner did not brood over this slight like some outcast ape, alone and abashed in his treetop. (He wasn't alone at all. At that point, he still had Jane Fonda draped on his arm.) He went on the attack, describing Murdoch under oath as a "slimy character" who "shamelessly" used the power of the media to advance his own wealth and influence. Nor was name-calling the only tool of aggressive dominance that came to hand. Turner was still brooding about Murdoch in June 1997, when he challenged his adversary to a pay-per-view boxing match. "It would be like 'Rocky,' kind of only for old guys," Turner joked. "If he wants, he can wear head gear. I won't." Murdoch's spokesman replied with a dour "no comment."

Aggressive dominance behavior clearly wasn't getting Turner the sort of traction he was seeking. In September of that year, he came up with an ingenious new strategy: Any damned fool can compete at the standard Darwinian game of gathering market share and piling up resources. But by giving away $1 billion to the United Nations, Turner could lay claim to the largest single act of charity by a living person in history. It was, of course, also an act of prosocial dominance, a bid for status, as plain as the chest-thumping of rival silverback gorillas or the philanthropic trilling of an Arabian babbler. Lest anyone mistake this for mere philanthropy, Turner characteristically announced the gift with hooting gibes at less generous billionaires: "There's a lot of people who are awash in money they don't know what to do with," he said. "It doesn't do any good if you don't know what to do with it." He mentioned Bill Gates of Microsoft by name, but, curiously, not Murdoch. Murdoch's limited public philanthropy to date had gone mainly to the Cato Institute and other conservative causes not dear to Turner's heart. Better to leave him awash in his billions.

Turner's criticism hit home among his fellow plutocrats. *Slate*, Microsoft's online magazine, responded by launching its *Slate 60*, an annual list of the biggest givers in the country. More important, *Slate* boss Bill Gates devoted much of the next year to reorganizing his own philanthropic giving. Until then, Gates had been known mainly for spending his money on a monstrous new Boy Wonder automated house, which critics had likened to William Randolph Hearst's San Simeon. Turner was by no means the first to suggest that Gates's previous charitable efforts, in the $100 to $200 million range, were cheap for a man with a net worth then estimated at $35 billion. But no one had illuminated the stinginess of the Microsoft founder with Turner's singular brio. (The dazzling scale of his gift encouraged people to forget that Turner himself was planning to fulfill his pledge over ten years, in installments of just $100 million per year.) "Well, I think Ted is great," Gates told Barbara Walters, during an interview on the ABC news show *20/20* a few months later. "And I'm very glad he has given that billion dollars. Certainly my giving will be in the same league as Ted's, *and beyond*." You did not need the italics added to hear the competitive challenge being taken.

Gates subsequently donated $1 billion to fund scholarships for minority students. The announcement was timed, according to a spokesman, "to let high school students apply for the scholarships for the next year." But the September 1998 announcement smacked of a different sort of deadline: It was just two days shy of the first anniversary of Turner's gift. A few months later, Gates and his wife Melinda announced additional gifts of $3.35 billion, followed by another $5 billion, and so on until they had created the largest foundation on Earth, and then some.

Slate, in an article by Robert Wright, was moved to liken the boss's "competitive generosity" to the potlatches of the Pacific Coast Indians who also once lived in Seattle. The idea of the potlatch was to lay on so much lavish generosity as to inspire awe in friends and silence in rivals. Or as the anthropologists Timothy Earle and Allen Johnson put it: "The Big Man and his following seek 'to flatten' the name of another group by 'burying' it beneath piles of gifts." One could almost hear the *Slate* tribe chanting, in the manner of the Kwakiutl: "*Our chief brings shame to the faces. / Our chief brings jealousy to the faces. / Our chief makes people cover their faces by what he is continually doing in this world, / Giving again and again oil feasts to all the tribes.*" To which the chief replies: "*I am the only great tree, I am the chief! / I am the only great tree, I am the chief! / You are subordinates, tribes.*"

But if Turner had been thoroughly flattened, how did his billion-dollar gesture constitute a dominance display? Was he better off for having done it? Or worse?

Turner had of course earned himself immeasurable goodwill by his gift, mainly because of the extravagance and the apparent selflessness of the gesture. Gates could do nothing to approach it, and Murdoch did not try. Even when Gates had given away $23 billion, roughly a third of his

2001 net worth (whereas Turner was still working on a third of what he'd had in 1997), critics saw it as merely a bid to influence public opinion and buy his way out of Microsoft's antitrust problems. Gates was running up against an unwritten law of successful philanthropy: The goodwill earned by a charitable gift is in inverse proportion to the donor's immediate need for it. Call it the rule of the decent interval. That is, you must establish yourself as a philanthropist for a decent interval of apparently selfless giving before you can cash in on the good name this earns you. (Sean "P. Diddy" Combs had the same problem when he tried to donate $50,000 to 100 Black Men, a prestigious think tank, shortly before he went on trial for the nightclub shooting of three people. The group rejected the gift as self-serving. Combs got off anyway.)

Turner meanwhile cashed in magnificently. While Microsoft was still floundering through its antitrust debacle, its competitor, the Internet provider AOL, announced plans to buy Time Warner. This deal represented a massive consolidation of Internet, cable television, film, and publishing interests. But it waltzed through its antitrust review, creating the biggest rival of the Internet age to Gates and Microsoft. And Ted Turner was the chief individual beneficiary. (AOL Time Warner stock has recently suffered a severe downturn. But the benefits consequent on Turner's philanthropy were manifold. Among other things, Turner was surely aware that, by inducing Gates to part with $23 billion, he had considerably narrowed the gap between himself and the richest man in the world. His UN gift had made him relatively less deprived. However enthusiastically Turner might disparage it, The *Forbes* 400 list was his Super Bowl, too.)

What about Murdoch? The ill will between the two men rankled long afterward. In 1998, a year-and-a-half after their dispute became public, Turner attended his first baseball owners' meeting in nine years to cast a vote against Murdoch's successful bid to buy the Los Angeles Dodgers. (Another case of Rupert wanting a baby just like Ted's.) In 2001, after Time Warner had become AOL Time Warner and Turner had been eased a bit further down the corridors of power, Murdoch unctuously allowed that he'd always liked his old rival. Missed him, even. "But now that he's out of work . . . I feel very sorry for him . . . when you read interviews with him and you can see his frustrations, you can't help but feel very sorry for him." This warm sense of fellow-feeling was probably the reason Murdoch's *New York Post*, at about the same time, ran the heartfelt headline: "Ted Man Walking."

Turner was still the single largest individual stockholder at AOL Time Warner. Moreover, his wealth had increased since his UN gift, just as he had predicted. The AOL merger alone gave him a one-day, $3 billion boost. "When I cast my vote for 100 million shares," Turner declared, with his usual lunatic gusto, "I did it with as much excitement as I felt the first time I made love some 42 years ago." By mid 2001, he was worth an estimated $9 billion, which, all things considered, was not a bad way to be out of work, especially as Murdoch himself was then worth about a billion less.

With his UN gift, Turner had also managed to make himself a hero not just at home in the United States, but in the fight against epidemic diseases around the world. Murdoch meanwhile gave $10 million to build a new Catholic cathedral in Los Angeles and got himself designated a Papal Knight. Otherwise, he did little to correct his reputation as a gnome, an ogre, a gargoyle. One article defending Murdoch began with this unpromising sentence: "It is a truth universally acknowledged that Rupert Murdoch is scum." It went on to say: "You may not like Murdoch—it's nearly impossible to like Murdoch—but you should probably admire him." The idea of defending Rupert Murdoch was clearly something you had to back into. Just as goodwill had its immeasurable benefits, ill will apparently had its costs. When Murdoch's News Corp. bid to purchase DirecTV, the satellite television service, Sen. John McCain suggested that his commerce committee might want to investigate "a consolidation of power the likes of which this country has not seen since William Randolph Hearst." Murdoch was reduced to sputtering, "I can't even believe you said this because look at AOL Time Warner . . ."

Correct me if I am mistaken, but the sound I seem to hear in the background is Ted Turner laughing every time he empties his bank account of another $100 million installment to the United Nations. From his perspective, there can be only one thing wrong with the whole picture: Unlike an Arabian babbler, Turner did not get the pleasure of actually stuffing his gift down Rupert Murdoch's unwilling throat.

6

The Service Heart

Subordinate Behaviors

I am his Highness' dog at Kew;
Pray tell me, sir, whose dog are you?

—ALEXANDER POPE, *engraved on the collar of a dog*
presented to Frederick, Prince of Wales

ONE TIME, WHILE VISITING A WOMAN AT HER APARTMENT ON PARK
Avenue, I asked who'd done her very elegant interior design, and she said,
"I did." Then, preening, she added, "Everything I do is about sensuality,
sexuality, feeling good." She pointed to a print on her meditation room
wall, showing a woman with her hand resting across a man's thigh.
"That's Samson and Delilah," she said. "My husband and I like to think
of that as us." A cloud of doubt crossed her face. "Not the haircutting and
all that." The good stuff. She was a shopkeeper's daughter from an outer
borough and rightly proud of how far she'd come. She described her technique for intellectual and spiritual self-improvement: "I gleam. I gleam
from what I read or hear."

"You really designed all this yourself?" I asked.

"Yes."

"Without anyone to help?"

Reluctantly, she said, "I hired a very nice man who was able to edit what I thought."

The very nice man turned out to be one of the most celebrated interior designers of the past quarter-century. His idea of editing was to say things like, "You must buy this, darling, or I will set my hair on fire." Once, on assignment for *Architectural Digest*, I'd asked him about a client's creative contribution, and he just rolled his eyes. "Hiring an interior designer," he said, "is like going to the dentist: You open your mouth and hope he does the right thing."

He did not, of course, say any such thing to the client, nor did I even for a moment contemplate using this quote in the magazine. Both of us understood our respective roles in the scheme of things. His was to make his wealthy clients feel as if their brilliant new teeth were in fact their own. Mine was to say, "Oh, what a lovely smile you have."

The first night she and her husband stayed in their new apartment, the Park Avenue client was saying, they lay in bed playing with the electronic controls for the shades, the music, the lights. Setting the mood. "It was very romantic. We'd arrived. In our home." Afterwards, they phoned up the designer. "You know how, after you have sex, you have a cigarette? I mean people who smoke? We put the phone in the middle and called Bobby and said, 'You're our cigarette.'"

Bobby was there, that is, to supply atmosphere, then vanish in a puff of smoke. And happy to oblige. People build brilliant careers, whole industries, even civilizations, on the slender premise of serving the rich with tact. With discretion. With minimal necessary rolling of eyes, however bizarre the whim. If the maharaja of Benares liked to be awakened each morning by a cow gently lowing at his window, and if business required him to visit a friend who thoughtlessly put him in a second-floor bedroom, then what was a faithful servant to do? Strap a cow into a harness, of course, and haul her skyward by pulley, to low for all she was worth at the maharaja's window. Naturally. Likewise modern rich people have a small army of personal assistants, makeup artists, trainers, nannies, interior designers, art buyers, investment advisers, publicity flacks, flattering journalists, bodyguards, friends and flunkies. And within every rib cage beats that splendid thing, the service heart, the heart that lives to supply the rich person's every want.

The service heart is of course far more complex and perhaps also more treacherous than it likes to appear. A Palm Beach hanger-on, a charity-ball walker, explained the subtext of the service heart this way: "The bad hustler says, 'What can I get from this?' The good hustler says, 'What can I give these people so they'll want me around for the rest of their lives?' " In terms of animal behavior, the service heart is perfectly natural. The friends and faithful servants who attend the rich display the same behaviors as do beta chimps biding their time and jockeying for position in the shadow of their alphas: We give our alphas attention, flattery, imitation, and anticipation of their every whim. We form coalitions to build them up (and also to repress, or at least commiserate about, their wretched excesses). We tolerate their abuse. We serve them with astonishing self-sacrifice, sometimes to the point of death. We use them. In time, we may betray them. But we are getting ahead of the story. Let's start, exalted reader, in the more promising land of flattery.

Whispering Sweet Nothings

A well-known New York architecture critic was recently quoted in *W* on the monumental aesthetic achievements of another Park Avenue matron, Courtney Sale Ross. "Courtney has one of the most extraordinary eyes I've ever encountered," the critic began, modestly. Then he cast off the gloomy pall of journalistic reserve and rose in a sort of exaltation to the task at hand: "She has possibly better taste than anyone, in terms of art, architecture, furniture, graphics. . . . She has pretty much infallible instincts across a wide range of different fields for real quality."

The readers of *W* must have wondered momentarily why they hadn't heard more about this towering genius. The answer was that Ross is neither a great artist nor the pontiff of some bold new aesthetic movement. She is merely, as *W* put it, "America's richest widow," with a $700 million fortune inherited from the late parking-lot-magnate-turned-Time-Warner-entertainment mogul Steve Ross. What the architecture critic was doing looked a lot like shameless sucking-up. It was the equivalent of a subordinate baboon delicately lipping burrs from the hindquarters of a dominant female.

Indeed, literally the equivalent. The word flattery comes from the Old French *flater* meaning "to stroke or caress." In his book *Grooming, Gossip, and the Evolution of Language*, British psychologist Robin Dunbar suggests that humans developed language (and also flattery) as a substitute for conventional primate grooming behaviors. Burr-lipping, louse-picking, and other highly personal forms of grooming serve, for most primates, as the main social pastime binding friends and family together, and also as one of the chief tools for social climbing. Monkeys and apes devote far more time to these activities than hygiene alone could possibly warrant, up to 20 percent of their day in some species. Grooming releases enkephalins and endorphins, the body's natural opiates, and the recipient blisses out on a tide of mild euphoria. The idea is that the grateful recipient will return the favor, assuming he or she has not fallen asleep.

Grooming of course has a tranquilizing effect on humans, too. But it doesn't translate terribly well as a social lubricant. A fifteenth-century courtier once discreetly picked a louse off King Louis XI of France, and the king graciously remarked that lice remind even royalty that they are human. Next day, an imitator pretended to find a flea on the king, who was by then perhaps tired of being human. "*What!*" he snapped. "Do you take me for a dog, that I should be running with fleas? Get out of my sight!"

Grooming with kind words is generally less intrusive. It's also far more efficient. Most primates live in groups of no more than forty or fifty animals, whereas humans are built, according to Dunbar, to maintain a reasonably close personal relationship with about 150 people. This is an awful lot of hindquarters to tend, or—if you happen to be a Courtney Sale Ross—an awful lot of people lining up eager to do the tending. Dunbar writes: "If modern humans tried to use grooming as the sole means of reinforcing their social bonds, as other primates do . . . we would have to devote about 40 percent of our day to mutual mauling." It is far more practical for all concerned to do it through the medium of *W*.

In common with monkeys and apes, we long to do our burr-lipping with the big boys and girls. That is, flattery and other forms of grooming are most rewarding when we are careful to groom the individuals at the top of the social hierarchy. In one study in Kenya, low-ranking vervet monkeys sometimes groomed a dominant individual for months or even years without recompense. If a subordinate was lucky enough to form an alliance with an alpha female, the subordinate would typically do the grooming ten times for every one time the alpha replied in kind. The phrase is, after all, sucking *up*, not down.

Monkeys, like humans, are extraordinarily strategic in their burr-lipping. At one point in the vervet study, a female named Marcos was on her way up the social ladder. Other monkeys were shrewd enough to foresee her rise and they gave her a disproportionate share of the grooming. In like manner, courtiers of King Louis XIV used to rise not just in the presence of his mistresses, but for any woman deemed likely to become his mistress. More recently, cosmonauts with an eye to attracting future space tourists were quick to note the salutary effects of zero gravity when California millionaire Dennis Tito arrived at the international space station. The mission commander told Tito, who had paid $20 million to get there, that he looked "maybe ten years younger now." Then he added, "we are going to prepare everything for you—nice bed and warm food."

Unlike monkeys, we humans can at least rationalize our more flattering and outrageous fictions with the thought that we are in good company. Almost everyone flatters the rich. For instance, John Milton, that immaculate Puritan, wrote his masque *Comus* on commission from the Earl of Bridgwater, a wealthy aristocrat eager to establish his family's reputation for sexual probity. When the masque was staged at Ludlow Castle, Lady Alice Egerton, the eldest daughter of the family, actually played the part Milton wrote for her. Faced with an enchanter tempting her into an illicit liaison, she delivered the immortal lines, "Mercy guard me! Hence with thy brew'd inchantments, foul deceaver . . . that which is not good is not delicious to a well-govern'd and wise appetite." This was, alack, a whitewash. Fifteen-year-old Lady Alice was no doubt entirely innocent. But her close kin had recently played the central parts in one of the more notorious sex scandals of the seventeenth century. Alice's uncle, Lord Castlehaven not only had sex with his household staff, a more or less accepted practice, but also compelled his wife, Alice's aunt, to have sex with them while he watched. He also collaborated in the rape of his twelve-year-old stepdaughter. The Crown ultimately lopped his head off on Tower Hill, the only point in Castlehaven's life at which his appetites might honestly be described as having been "well-govern'd."

· The rich themselves seldom seem to notice that flattery is everywhere in their lives. "If someone is trying to flatter me, I walk away," says Mark Cuban, a billionaire who sold his Internet company before the dot-com bust and now owns the Dallas Mavericks basketball team. "If someone is dishing b.s. at me, I usually laugh and get out of there. I've been grabbed, cornered, hugged, squeezed, but it's pretty much meaningless to me." Cuban communicates mainly by E-mail now, to avoid being grabbed, cornered, hugged, squeezed, or flattered. But a few paragraphs later in the same E-mail he writes, "I especially love Shaq. He always comes up to me and smiles and tells me I'm his hero and that he wants to be like me IF he grows up." It may be that the rich don't notice how important flattery is in their lives because it is so routine, so institutionalized. Ideally, it is also invisible. In Imperial India, for example, wealthy and powerful visitors seldom shot a tiger of less than ten feet in length, in part because their *shakiri* shrewdly used tape measures with an eleven-inch foot, "so that . . . a ten foot tiger automatically became an 11 foot tiger in front of every-body and to everybody's satisfaction." Likewise, in twenty-first-century Los Angeles, personal assistants buy Levi's for the boss, then send them out to a business called "Ragtime Denim Doctors" to have the waist-size label replaced with one from a slimmer pair of jeans. The tendency to believe pleasing things about oneself runs deeper in the rich than in the rest of us because we supply them with physical evidence for believing. We want them to believe that the waist-size label is real, that a nice man merely edited their design ideas, and that the infallible aesthetic instincts belong to them and not their hirelings. We do it because, in the long run, stroking the egos and serving the whims of the rich is good for us.

Whim-Tending

Derrell Harris was born to serve the rich. As a thirteen-year-old in Amar-illo, Texas, he had charge accounts at three different florists. "I'm a giver," he says. "I love to give. I like to make people feel that they are loved." So much so that his sister had to pawn her stereo to help pay off the bills. "My heart was *so* big. My budget wasn't big. That's who I am." Harris started out as a caterer, but now works twelve-hour days as a personal assistant for a wealthy employer in New York. He earns $80,000 a year, plus living expenses and his own studio apartment (no view) in a prime Manhattan neighborhood, plus frequent flyer miles, plus-plus-plus, for Christmas one year, a $7,000 David Yurman watch. "I was like, *wow*, so it does pay to serve."

His budget is now almost as big as his heart, and Harris exudes the personal assistant's characteristic need to please. "If somebody asks for it," he says, "I feel that it should be delivered immediately." His boss once asked him to convert another studio apartment in the building into a workout room on a week's notice. Harris found contractors (yes, in Manhattan) who had the place gutted, replastered, repainted, recarpeted, with mirrors on the walls and the gym equipment in place five days later. "There are times when I just kill myself to get it done. What did it cost? Oh, I can't disclose that. A lot. It was great. He was all smiles. Whether he didn't use it for three months doesn't matter. He asked for it."

Harris learned his craft at Starkey International, a training and placement agency in Denver, where they espouse the idea of the service heart. "People with a service heart will go the distance, they'll do what the principal needs," says Bill Bennett, who runs the training program there. The holy mission at Starkey is to professionalize household service. Given the budget, staff, and multiple homes of the average wealthy family, Bennett suggests, the job is equivalent to running a small company. So Starkey always talks about "household managers." The term "personal assistant" apparently originated in Hollywood and still carries an unfortunate hint of glitz (when what one wants, my dear, is patina). And the word "butler" rings false in a country where Jeeves is best known as an Internet search engine. It conveys the unpalatable suggestion that the individual is a servant. "The butler," says Bennett, "is there as a figurehead, to answer the door. Household managers are down-and-dirty. They're commandos: Get in and get the job done."

By any name, the personal assistant's job, bordering at times on sacred calling, is to take care of life's tiresome little details—the grocery shopping, the flight plans, construction of the new house in the Hamptons, the phone call from Mom ("He hasn't called in a week, I could be dead!")—so the boss has time to focus on finer things. Or as the owner of one placement service put it: "All the stuff that you and I have to do for ourselves, and wreck our weekends, these people do."

Downstairs in the Starkey building, a 1901 Georgian Revival mansion, Bennett is leading a half-dozen household managers through basic training. They're a mixed group, by age, race, and gender but all are dressed in mansion-commando attire—blue blazers, khaki slacks, and brown leather shoes. Most of them already have jobs with rich clients, who have sent them here to brush up their skills. They volunteer the sorts of things their bosses routinely expect them to get done: "Check and replace light bulbs daily," says one.

"Oooh," says Bennett. "Big one. No burned-out bulbs. Major, major, major." And if the boss unscrews a bulb just to test your diligence? "You screw it back in and go up a notch." "Vacuum in a straight line," says another student.

"And probably when you do that," says Bennett, "you back out of the room vacuuming. No footprints."

What about food? "Fresh," says a student. "Fresh Starbucks," says another, and then, rapid-fire, "Organic," "Spa cuisine," and "Low fat," until a student, the wicked angel of culinary indulgence, cries, "Fat equals flavor. Cream sauce!"

"The more information you can get on how they want it done, the better," says Bennett, "and then write it down and don't deviate, because if you deviate they'll wonder whose agenda are you on."

Ziplessness

The idea is to make the household function flawlessly, invisibly, as if by magic. Or as David Athey, a personal assistant in New York puts it, the employer "should never have to ask for the toothpaste. That should just be there. He should never have to ask for toilet paper. I know he likes Diet Coke, so I'm sure to have it in the refrigerator." A former employer with homes in London and New York once told him, "I don't ever want to have to walk into my home and think about it." What's wanted isn't Erica Jong's Zipless Fuck (though that would be fine, too) but the Zipless Life.

Not having to think about it is of course liberating. It can also be disabling. One personal assistant got a call at 4:00 A.M. at home in New York from her boss who'd run out of toilet paper in his London hotel and wanted her to fix it. This is an odd twist on history: The old aristocracy in places such as England and India grew up with servants and practiced being incompetent from birth. Now the aristocrats are out in the market-place desperately trying to prove they can cut it as meritocrats. The mer-itocrats meanwhile need to display real skill to get to the top, where they then achieve helplessness as the reward for former know-how. "They used to be able to pour themselves a cup of coffee," says one assistant. "But when you arrive at 7:30, they're not able to do that any more. Whatever they need, you're there to take care of." Her urgent chores once included packing Nathan's Hot Dogs in dry ice and overnighting them to Lyford Cay in the Bahamas, where the boss thought they'd make an amusing vacation dinner.

Deploying a personal SWAT team in hot pursuit of some whim helps validate the rich person's importance, and in a sense, the more trivial the mission, the better. LeeAnn Heck, who now runs a Los Angeles business called Consider It Done, used to be a personal assistant to a leading actress, who developed a life-or-death need for Irish butter, whipped, not salted. The dietitian said she could not eat her potato without it. "Five people worked on it for three days," says Heck. "We called every place from San Diego to Santa Barbara. It became a real point. '*Where's the but-ter?*' It became like checking on a sick pet or a child. This is somebody who has so much to do. It was amazing that it took over her life."

Another personal assistant's boss wanted seedless organic raisins. "They looked like pennies that had been run over on the railroad track," says the assistant, who found them after phoning every major health food store in the nation. "I put them in the pantry, and they went untouched for two months. The day they're opened, the client hadn't had his break-fast and he had low blood sugar, so he's in a bad mood to start with. So I hear my name. 'These are the worst fucking raisins I've ever eaten,' and he threw the raisins at me. He threw the whole pack."

Yet the subordinates of the rich often clearly enjoy their status. They sometimes become subsumed in their employer's identity and find satis-faction only in meeting the employer's needs, like Mrs. Danvers in the Daphne du Maurier novel *Rebecca*. "When I go to bed at night," says the guy who got pelted with seedless organic raisins, "my last thought is, What have I done today to make your life better? And my first thought in the morning is, What can I do to make your life easier today? And the only thing I want in return is to be treated kindly and humanely." The comforting, but vaguely creepy thing, is that he sincerely means it.

Powerful Attraction

So why do subordinates put up with it? Why do we indulge rich people in their unreasonable demands, suffer their arrogance, repay their generosity with the kind of loyalty we might otherwise reserve for our own families, and even honor them as citizens of the year, patrons of the arts, friends of the Earth, humanitarians? Why do even spouses and children of the rich allow their kin to keep them in a debilitating state of dependence and uncertainty? The short answer is that we do it for the money.

But this isn't an adequate answer, if only because, for the most part, we're not going to get it. So here's the long answer. We crave social hierarchy in part on account of fear and the need for protection, and it's a craving deeply rooted in primate evolution and also recapitulated in every small child's life. "Higher vertebrates initially seek refuge with their mothers," writes Irenäus Eibl-Eibesfeldt, the Austrian ethologist and anthropologist. "This is as true of chickens as it is of people." The powerful attraction to individuals who are dominant, larger than life, gradually shifts to individuals outside the family. "In baboons the young animal always turns to its mother during the first few months of its life, but later on looks to other high-ranking adults for protection. The most common goal-in-flight is the highest-ranking male, even—interestingly enough—when it is he that has elicited the flight response." The gravitational pull of the dominant individual may be one reason battered spouses and abused children cling to the person who mistreats them. For instance, Mary Lea Johnson, a Johnson & Johnson heiress and the first baby to appear on the company's baby powder label, claimed that she was sexually abused beginning at the age of nine by her father J. Seward Johnson, but she remained devoted to him until his death. Among wealthy families, the attraction of the alpha may be why later generations remain fiercely loyal to the image of the founding father who still dominates them, though in private they bitterly resent the feeling that their lives are not their own.

Having a common enemy outside the group also encourages bonding, so much so that subordinates frequently seek to shine in their alpha's eyes by attacking his enemies, like the minions of King Henry II rushing out to slaughter Thomas à Becket. Or, a bit more prosaically, like the producers of a recent sitcom called The Chimp Channel, which aired on Ted Turner's TBS. Apes played all the parts in a behind-the-scenes farce at a cable television network, run by a media baron originally conceived as "a nutbar southern billionaire." Nobody had to tell the producers to do a rethink, and they wisely transformed the media baron into a stupid, mean-spirited Australian. The enemy of the rich man is my enemy (at least until he cancels my series). Another Turner subordinate has published a book titled, *Me and Ted Against the World*.

Fitting into the hierarchy gives us a sense of security, and so we get special pleasure from serving the people at the top. "We have a kind of childlike view of rich people. We regard them as people who can take care of us," says Peter White, who serves as a sort of priest to the wealthy in his position as managing director of the family advisory practice for Citibank. "When I go to wealth conferences, I am constantly hearing what a wonderful person A and B and C and D are. And I think it's sincere. I think there's something in all of us that wants to believe that rich people are so wonderful, just as when we are children we all want to believe that our parents are so wonderful." This filial devotion is so powerful that at Drexel Burnham Lambert in the 1980s, a salesman for Michael Milken started calling the junk bond king "Dad" and once pasted his own picture into a family photo on Milken's desk.

The need for protection is only the most primal source of our attraction to dominant individuals. Hierarchy obviously also has more practical benefits. At least in theory, a hierarchy allows individuals within a group to function more effectively. It's why committees have chairpersons, so they can concentrate on the task at hand. The alternative is to squabble endlessly about who ought to be in charge. In one study, researchers left the pecking order in some chicken flocks undisturbed, but they deliberately unsettled other flocks week after week by removing whichever bird had struggled to the top. The flocks with undisturbed hierarchies did less bickering, and even the subordinates ate more food, gained weight faster, and produced more eggs.

Sorting out rank in the first place can be perilous for everyone. So once the hierarchy is settled, it pays to avoid further bloodshed and instead acknowledge rank with ritualized gestures of dominance or submission: The rich man preens and tells bad jokes. His subordinates gather around and laugh appreciatively. The chauffeur tips his hat.

Underlings often find comfort in their status, a reassuring sense of boundaries and limits. Think of the gardener's horror, in the rigid old British class system, against any attempt to rise above one's station. Regarding his rich employer as almost another species may have helped the gardener to spend a lifetime weeding the great man's footpaths without suffering an affront to his own self-worth. We may also enjoy being subordinates because the rich and powerful give us the thrill of association, a vicarious sense of glamour and importance. A bit like a dog with its master, its hind end in a swivet before it flops on its back in a pure gleeful grovel.

In the animal world, subordinates often declare their status with an air of delighted submission. Among the chimpanzees Frans de Waal studied at a zoo in Arnhem, The Netherlands, submission displays had the character of courtly ritual: "The subordinate assumes a position whereby he looks up at the individual he is greeting. In most cases he makes a series of deep bows which are repeated so quickly one after the other that this action is known as bobbing. Sometimes the 'greeters' bring objects with them (a leaf, a stick), stretch out a hand to their superior or kiss his feet, neck or chest. The dominant chimpanzee reacts to this 'greeting' by stretching himself up to a greater height and making his hair stand on end. The result is a marked contrast between the two apes, even if they are in reality the same size. The one almost grovels in the dust, the other regally receives the 'greeting.' " De Waal could be describing supplicants in the court of Louis XIV.

Among humans, oddly, even the rich and powerful sometimes display the subordinate's love of hierarchy. Culture makes us aware of other hierarchies, some of them more venerable or more glamorous than our own. A manufacturer may be king in the world of prosthetic devices, say, but if Warren Buffett comes to town the limb king greets the value-investing king with a subordinate's deference and he comes away with a subordinate's vicarious sense of importance. Nothing like this exists in the animal world. No silverback gorilla pines for acknowledgment by that very important silverback over on the other side of the Virungas. But the rich do. Nancy Mitford, the daughter of a baron, observed the folly of her class at firsthand. In her novel *Love in a Cold Climate*, Mitford described the animal delight with which the otherwise haughty Lady Montdore does obeisance to visiting royalty: "Her curtsies, owing to the solid quality of her frame, did not recall the graceful movement of wheat before the wind. She scrambled down like a camel, rising again backside foremost like a cow, a strange performance painful it might be supposed to the performer, the expression on whose face, however, belied this thought. Her knees cracked like revolver shots but her smile was heavenly." It was blissikins for Lady Montdore even to breathe the same air as such exalted company.

Just Normal People

The cumulative effect of all this running and fetching and bowing and scraping can be profound, and this brings us back to the commonplace assertion by the rich that they are "just normal people." "Wealth doesn't change people," Henry Nicholas of Broadcom has remarked. "It changes the way people treat you." But the truth is that nothing changes an individual more than the way other people treat you. The deference of others may actually alter the neurochemistry of the individual's brain. Serotonin is a neurotransmitter, a substance that eases the flow of nerve impulses across the synaptic gap, and it seems to make people more relaxed and socially assertive. When increased by drugs such as Prozac, it also helps to fight depression. Serotonin levels normally fluctuate according to our everyday experiences. In one study of vervet monkeys, high-ranking individuals had serotonin levels twice those of their underlings—and the bowing and scraping of the underlings was apparently an important reason. The research team, led by Michael T. McGuire of the UCLA Medical School, placed an alpha vervet in a room with a one-way mirror, so the alpha could see his subordinates, but they couldn't see him or respond to his dominance displays. Over a period of several days his serotonin level plummeted. His sense of well-being apparently depended at least partly on receiving his minimum daily dose of deference and submission. The chance to make their submissive displays was clearly important for the subordinates, too. Elevated serotonin actually seemed to inhibit destructive aggression in the dominant vervet. It enhanced his skill at gaining allies and other means of prosocial dominance.

McGuire subsequently suggested a similar connection between social dominance and serotonin levels in humans. In one informal study, never published in a scientific journal, he found that officers in a college fraternity typically had more serotonin than their less powerful frat brothers. They weren't apparently born that way, destined at birth to be president of Delta Kappa Epsilon. It wasn't the elevated serotonin that made them dominant. Rather, dominance seemed to give them elevated serotonin after the fact. In a sense, the deference of subordinates gave them the tools for leadership once they got the job.

For the very rich, the deference of subordinates can be profound, even abject. The former head of Vivendi Universal, the media and utilities group, is known among his French employees simply as J6M, for Jean-Marie Messier, *moi même, maître du monde*. Henry Fok, the Hong Kong mogul who likes to travel in a private 747 (with a second private 747 trailing behind for his staff) is never referred to by name but only as "The Leader." Deference by itself doesn't of course guarantee an individual's well-being. Ted Turner, for instance, is well-known as a manic-depressive. But by and large all this deference seems to be a good thing, biochemically speaking, for the rich.

Stressed-out Subordinates

In some animal species, the social hierarchy is so important that it puts its mark on the flesh-and-blood physiology of dominants and subordinates alike. Mice, for example, can literally smell dominance. They can identify the scent-markings of a male who has previously defeated them. Oddly, they tend not to retreat from the dominant male's scent but to creep tentatively closer on flattened belly, as if wondering, Is the alpha mouse that whomped me near enough to do it again? and If he's near enough, maybe cringing will dissuade him from further whomping.

Is there any human equivalent? Earlier in this book, I wrote that the rich are different from you and me; they do more scent-marking. It was of course a joke. Irving Berlin had the same idea when he wrote his song, "Slumming on Park Avenue," about the joy of sniffing around the way the snooty sorts do, but on *their* territory, not ours. Berlin did not quite suggest that rich folk spray the soles of their feet with urine, as voles and bush babies do, the better to define the boundaries of Upper East Side chic with every footstep. But the legendary "smell of money" was certainly a subtext, and one that's pervasive in writing about the rich. In his recent novel *Turn of the Century*, for instance, the critic and erstwhile publishing minibaron Kurt Andersen describes a media mogul as smelling of "the daily haircut plus fresh flowers plus cashmere plus BMW leather plus the executive-jet oxygen mix plus a dash of citrus. That is, [he] smells luscious. He smells rich." This scent has a profound effect on an otherwise jaded subordinate, rendering him "suddenly as taut and overwrought as a teenager," given to piping his agreement with whatever the boss suggests. It makes him, in a word, submissive—like a beta mouse.

Because of the diminished sense of olfaction in our species, it would be absurd to suggest that humans can routinely smell social dominance, much less wealth. But dominance gets under our skin in other ways, as evidenced most notably by the hormone cortisol. A product of the adrenal cortex, cortisol has a powerful influence on mood, blood pressure, muscle formation, immune cells, inflammation, and gastrointestinal function. A spike in cortisol levels is also essential for dealing with serious physical or mental stress. In dominant individuals, cortisol tends to spike powerfully in the face of a threat, then quickly drop back to a relatively low resting level. Individuals with this profile are typically adept at ignoring false alarms and aggressive at slapping down the real threats. Subordinate humans, like subordinate mice, typically have the opposite profile: Chronically stress-heightened cortisol levels at rest and not much of a spike in time of need. False alarms make them skittish, and they don't have what it takes to put down a serious threat.

The reality is no doubt much subtler than this. A recent study of baboons, for instance, found that there were at least two subclasses of subordinates: The skittish, downtrodden ones, and another subclass of ambitious subordinates destined to achieve high rank within five years. This second subclass displayed the sharp cortisol spike of dominant individuals. But they also suffered from the high resting cortisol levels of subordinates, apparently because of the stress of being in a transitional stage. Among humans, the picture is probably even more complex, depending, for instance, on whether one is a subordinate to a notoriously difficult boss like Miramax cochairman Harvey Weinstein (who once made an unsatisfactory assistant chant, "I'm a dildo, Harvey") or to some comparative sweetheart.

Chronically elevated cortisol levels can have serious physiologic consequences, including high blood pressure, lethargy, depression, and muscle loss (the rich get richer, the poor get weaker). Cortisol also influences the accumulation of abdominal fat. A recent study found that women who rated themselves as having low socioeconomic status not only had higher cortisol levels but, apparently as a result, also suffered the lower-class indignity of the tubby gut. Women who regarded themselves as rich and powerful had lower cortisol levels and tended to be slimmer and healthier.

Eunuchs

In many animal species, even more alarmingly, elevated cortisol levels lead to what biologists call "psychological castration." Subordinate male mice have depressed sperm levels. In platyfish, where sex roles are reversed and females compete for males, the alpha female inhibits ovarian activity in her subordinates. Among baboons, beta males learn that love literally hurts. They shy away from mating opportunities for fear that the dominant male will catch and punish them. Among gelada baboons, which are a separate species, a subordinate female suffers relatively little harassment by superiors. Even so, for each step down the social ladder, her lifetime reproduction rate falls by one-half an offspring.

No one has actually studied psychological castration in humans. Even amid the current efflorescence of evolutionary psychology, it would take a very intrepid young doctoral candidate to propose a thesis with the title, "Do mansion commandos suffer from depressed sperm levels?" or "Does nanny have cobwebs in the womb?" But the history of the relationship between rich people and their subordinates is suggestive.

In many civilizations until recent times, for instance, the nobility did not nurse their own children. Instead, they employed wet nurses, often poor country women who sacrificed their own infants to preserve their milk for the children of their aristocratic employers. In some cases, girls apparently became pregnant with the intent of getting rid of the child and selling their milk. Benjamin Haydon, the historical painter and drinking companion of the Romantic poets, recorded an appalling instance in the early nineteenth century. He and his wife had already lost five children, and their infant daughter was wasting away. Then he found a wet nurse living with her "fine fat pink baby" in dire poverty. Despite suffering torment about whether it was just "to risk the life of another child to save my darling Fanny," Haydon hired the wet nurse. Fanny "seized the bosom like a tigress and was saved," Haydon wrote in his diary, "but the fine baby sank and perished."

In imperial China, poor families frequently volunteered their sons, and adults sometimes offered themselves, to become eunuchs, "for a eunuch could be sure to obtain an easy and lucrative position in the Imperial Palace or a princely mansion." This was decidedly not psychological castration: "The operation performed on eunuchs was a crude one, both penis and scrotum being removed in one cut with a sharp knife," the China scholar R. H. Van Gulik has written. As recently as 1890, a hereditary castrator outside the palace performed the surgery for "a high fee, which could be paid in installments later when the persons operated upon had obtained a position in the Palace." Castration was one way to ensure that the mansion commandos remained on the principal's agenda, their own agendas having been somewhat curtailed.

Among modern plutocrats, mutilating household staff is of course no longer de rigueur. But remarkably often, the people who work in the homes of the rich choose to be single and childless. One longtime assistant remarked that she could not think of any personal assistant who has two children. Another assistant, who works for the television personality

Bryant Gumbel, promptly replied, "Is that including your boss?" The rich and powerful typically require so much tending that their subordinates can't find the time for a life of their own. Or as the assistant to Sean "P. Diddy" Combs has put it, "It's always a pull back and forth. She'll say, 'You're up in Puffy's ass.' And Puffy'll be like, 'You're all up in your girl's ass.'" They are on-call around the clock. "Have pen and paper handy at all times, even in bed," one assistant advises. They do not always know what city, or what country, they will be in on Saturday night. So dating gets awkward. If they happen to meet an attractive stranger, they often cannot say what they do for a living. "The minute they say they work for a celebrity," says a veteran of that world, "the person they're talking to wants to know about the celebrity, and not the assistant. So most of them say, 'I'm a secretary.'"

Celibacy is the inadvertent result, a product of circumstances and individual choice. But the rich employer typically creates those circumstances. Oddly, it's sometimes easier to see deliberate psychological castration within the rich person's own family. Because such families focus so intensely on preserving net worth and passing it on through inheritance, the question of who gets to reproduce can take on extraordinary importance. This is especially true when the intent is to pass on a family business or a family estate intact to just one heir. In her book *Passion and Prejudice: A Family Memoir*, about the prominent Kentucky family who owned the *Louisville Courier-Journal*, Sallie Bingham describes how her father became the principal heir in 1937 despite being the youngest of three children: His older brother Robert was a renegade who drank too much. When Robert contracted syphilis, "the Judge," as their father was known, "ordered the mercury treatment that left Robert sterile, treatment he had rejected in his own case." He also kept a sister Henrietta, who remained unmarried, "suspended in endless hope, endless excitement," ultimately leaving her only a small share of the family wealth. The Judge, writes Bingham, "had in effect neutered two of his three heirs— the two who might have questioned the family myth." He could afford to do so because his stable and less threatening youngest child had already produced two infant sons to carry the family enterprise into the future. Bingham adds a chilling Freudian postscript. When she was a child on the family estate, which her father had inherited, her mother sometimes assigned her to prune back "the sticky peony buds" in the garden: "The Darwinism of this project was, I realized later, peculiarly applicable to the family, in which lesser members were sometimes sacrificed to the growth and display of the central figures."

There is of course a significant difference between family and subordinates when it comes to psychological castration: Within the family, limiting reproduction can be an essential strategy, and the neutered siblings at least enjoy the perpetuation of their genes through their nieces and nephews. With employees, it is merely a convenience for the rich employer.

Good God, Man, Isn't That What We Have Servants For?

Let's, for a moment, consider an even creepier aspect of subordinate behavior. One night over dinner in a restaurant in Ciudad Juárez, Mexico, a businessman was telling me about a Middle Eastern potentate who owned a 747, with a gyroscopically rotating prayer room perpetually oriented toward Mecca. The potentate had a serious heart condition, the businessman continued, so he converted the plane's upper deck into a cardiac intensive-care unit with all the latest technology. I made suitable murmurings of awe. The businessman smiled patiently until he got to the part that had impressed even him: "The plane was also equipped," he said, "with a living donor." A heart donor, that is. It was a poor man of compatible tissue type whose reward, it seemed, was living well for a little while and the promise that his family would live well afterward.

The story naturally stuck with me. This was the service heart to the ultimate lub-dub, and I wondered if the real reward for the donor wasn't the almost sacramental privilege of becoming one with his potentate. I was eventually able to locate a private pilot who had actually been on the plane, visiting a fellow pilot. He confirmed the prayer room (in the nose, just under the cockpit) and the cardiac intensive-care unit, and he gave me the name of the potentate, King Khalid of Saudi Arabia. But, alas, he had not seen anyone onboard with a queasy smile and a splash of iodine over his heart. In any case, Khalid had died in 1982, at home, of a heart attack. So maybe it was just a story, a nice twist on the old Yiddish conundrum: If the rich could hire other people to die for them, the poor could make a wonderful living.

But a beta wolf will fight to the death for its pack, and soldier termites readily sacrifice themselves to defend the king and queen of their colony. So the 747 story left me wondering whether there are in fact circumstances in which our readiness to subordinate ourselves to the rich sometimes translates into a willingness to die for them. With the exception of bodyguards, who are, like soldier termites, a separate caste, the notion is repulsive to our democratic ideals. But the rich have, of course, often hired people to risk death in their stead. The young J. P. Morgan and John D. Rockefeller both hired substitutes to fight for them in the Civil War, as did almost everyone else who could afford the $300 price, roughly a year's wages for an industrial worker. Not content to have done his duty in this fashion, Rockefeller also once reached into his safe and handed a ten dollar bill to each of thirty new Union Army recruits who had been ceremonially marched into his office. "God, but he must be rich," one young soldier remarked, the contemporary equivalent of a Roman gladiator saying, We who are about to die salute you. Rockefeller was even richer five years later and, unlike many of the soldiers, still breathing.

The system of hiring substitutes was even more patently unjust in the South. The notorious "twenty nigger" law gave the largest plantation owners, with twenty or more slaves, an automatic exemption from military service. Lesser planters hired substitutes, who were in essence fighting to preserve a system that contrived to keep them poor and disenfranchised and to defend an aristocracy that regarded them, like the slaves themselves, as not quite human. Rebel soldiers routinely beefed about this in their letters home, according to writer Andrew Ward, but they nonetheless fought and died, and their families back home starved and suffered, "so the men in the Big House could waltz."

In times of crisis, the unspoken rule of many highly stratified societies was that the rich should live and the poor should die. In part this was a natural outcome of rich families having greater access to vital resources. But University of New Mexico researchers James L. Boone and Karen L. Kessler suggest that the poor often felt they had less "right to survive." On the tiny Polynesian island of Tikopia, for instance, a 1929 hurricane flattened coconut and breadfruit trees and ruined the taro and other staple root crops. The population of 1,281 people, separated by two hundred miles of open water from any substantial neighboring island, suffered a yearlong famine with up to three people starving to death each week. Ethnographers who visited the island soon afterward reported, "What was communicated explicitly here was that the chiefly families would be the last to die. . . . In fact, everyone, commoners included, felt that it would be unthinkable for the chiefs to die." There was a kind of magic about the upper classes: "Theoretically, the chiefly families own the land and there is a mystical connection between the well-being of the chiefs and the well-being of the land." In a similar catastrophe in the Caroline Islands in 1775, 900 people were killed outright by a typhoon and another 80 starved to death in the next few weeks, leaving only 20 survivors, among them the hereditary chief and members of his household. Even the supposedly egalitarian Hopi Indians of the American Southwest concentrated their resources, in times of drought, on the survival of high-ranking clans, forcing poor families "to migrate or starve."

The tendency of the upper classes to outsurvive other classes in a crisis, and for the poor to die in their place, may persist even in complex modern societies. The stories that stick most vividly in our minds from the sinking of the *Titanic*, for instance, are about rich people going honorably to their deaths: Benjamin Guggenheim refusing a lifejacket and dying dressed in evening clothes (insisting, by the way, that his young valet do the same). Mrs. Isidor Straus dying because she refused to step into a lifeboat without her husband. John Jacob Astor putting his pregnant eighteen-year-old wife aboard a lifeboat and promising to follow soon after. One survivor recalled: "As the excitement began I saw an officer of the *Titanic* shoot down two steerage passengers who were endeavoring to rush the lifeboats. I have learned since that twelve of the steerage passengers were shot altogether, one officer shooting down six. The first-cabin men and women behaved with great heroism."

But when Boone and Kessler analyzed survivorship, they found that 33 percent of adult males in first class survived the sinking versus 16 percent in steerage. The percentage of first-class adult males escaping was almost equal to the percentage of children escaping from steerage. Overall, 62 percent of people in first class survived, and only 25 percent of those in steerage. This may sound unsurprising. The rich had, after all, purchased better access to vital resources, the lifeboats, by paying for rooms on the upper decks. But what caught the attention of the researchers was that adult males in second class, though also closer to the lifeboats, died at a far greater rate than their counterparts in steerage. Boone theorizes that the rich survived because they could accept dishonor and go home to the sheltered retreat of their estates, and the poor were too far down the economic scale to see any point in dying for honor. But the middle-class males, being upwardly mobile and practiced in the art of deference to their social betters, would have regarded giving up their seats on a lifeboat as a necessary display of personal honor. Thus only 8 percent of them survived.

Still, honor doesn't count for much any more, and the idea of dying for the rich seems like a highly improbable form of subordinate behavior. Somewhere down below murdering them, say, or suing them for sexual harassment. So I was prepared to write off the idea of a living heart donor as apocryphal. Then, in November 2000, the story broke that Kerry Packer, a sixty-two-year-old billionaire who controls Publishing and Broadcasting Ltd., the second largest media empire in Australia, had been hospitalized for a kidney transplant. Packer had earned a reputation, in the words of his fellow publisher Conrad Black, as being "generous and a bully, brilliant though dyslexic, domineering but convivial, and fiercely possessive, ferocious and vindictive in dispute." Not necessarily the sort of person for whom people line up to shell out their internal organs, especially given Packer's unpromising medical history of cancer and heart trouble. But in fact Packer found a living donor in his own helicopter pilot. Nicholas Ross, a sixty-year-old veteran of the Royal Navy, had worked for Packer and been his close friend for almost twenty years. Packer, who referred to Ross fondly as "Biggles," had built a home for him on his property outside Sydney. Donating an organ was of course not the same thing as dying for the rich, though donors risk a one-in-five-hundred chance of suffering kidney failure themselves. But it was certainly "an extraordinary act of kindness and generosity," in the words of Packer's son James, who was not a donor. So why did he do it? Ross was one of the best-paid helicopter pilots in Australia, but the motive for his gift surely went much deeper. Packer was not merely his friend, but his hero, his provider, his protector. Despite the two-year age difference been them, the pilot in fact liked to address Packer as "Father."

Monkey See, Monkey Do

Well, OK, I see that I am in danger of caricaturing subordinates to the rich as a class of eunuchs employed to goose up the serotonin, drive the car, and, in times of need, supply spare body parts. Prudence at this point dictates some judicious backpedaling. But let's plunge one step forward instead: Subordinates often aspire to be the rich person's clone—figuratively speaking, that is. The urge to imitate the rich and powerful is almost irresistible, and, from the Darwinian point of view, it's a highly adaptive behavior. When a subordinate enters a new household, imitation is the usual first step toward fitting in and breaking down mistrust. Done subtly, imitation flatters the rich person and thus might conceivably help secure his favor. An early handbook for upwardly mobile Egyptians, *The Instructions of the Vizier Ptahhotep*, advised: "Laugh after he laughs, and it will be very pleasing to his heart."

Imitating our social betters typically means imitating success, rather than failure, and the habit is thus deeply ingrained in primate nature. For instance, Jane Goodall reports that after Mike the Chimp established himself as the alpha male by banging together two empty kerosene cans, researchers occasionally spotted a young male named Figan off by himself in the bushes practicing with Mike's discards. (Unfortunately, he was rehearsing his dominance displays with only one can, not two, the difference between *savoir faire* and imitation.) An experiment by the primatologist Robert M. Yerkes also suggests that chimps mainly copy high-ranking individuals. The Yerkes team removed a low-ranking individual from a captive troop of chimps and taught him complicated manipulations to extract bananas from a special feeding apparatus. None of the other chimps watched this rank upstart long enough to figure out for themselves how to extract the bananas. They just let him do the work, then walked up and swiped the reward. Later, the researchers taught the same complicated manipulations to the dominant individual in the troop. This time, everyone said, "Ooooh! Waaaah!" and promptly set out to imitate their ingenious alpha.

Subordinates assume that alpha knows best, and we imitate him almost reflexively. As Tevye suggests in "If I Were a Rich Man," from *Fiddler on the Roof,* it doesn't matter what the question is, or whether the answer is right or wrong, people just assume a rich man really knows. Among his chimps at the zoo in Arnhem, Frans de Waal once jotted down this note after the alpha named Luit suffered an injured hand in a fight and was forced to walk on his wrists for several days: "Amazingly, all the young apes imitate him and suddenly begin stumbling around on their wrists." One might think that no human subordinate would ape a superior in quite so mindless a fashion. On the contrary: In China, the wealthiest women used to have their feet bound so tightly that it rendered them permanently incapable of walking. Being crippled in this fashion was a status symbol, a mark of the ability to hire servants as bearers. Poor rural women couldn't afford to give up their natural mobility, of course—yet when visiting town they dressed their feet to *appear* bound. The urge to imitate was so powerful, certain outcast groups had to be legally barred from foot-binding.

On a somewhat grander scale, the Italian ambassador Primi Visconti recounted the slavish imitation among courtiers to Louis XIV: At one point when the king's sexual misbehavior had apparently become egregious, the duc Mazarin "told the King that he had had a revelation that night that His Majesty was to behave better; to which the King answered: 'Well, I dreamed that you were mad!' Immediately everyone, down to his own footmen, treated the duc as if he had been a madman so that he no longer dared show himself at court. Several years later, the duc . . . told the King how low he had fallen and begged for help." When his court came to greet the king in his bedroom one morning soon after, Louis made a point of talking about hunting with Mazarin, "then, turning to the courtiers, he said that the duc had wit. Hardly had the duc left the bedroom before more people crowded around him than around the king." Today we like to think we are beyond imitating kings. Yet we still wear vests with the bottom button left open, in imitation of King

Edward VII, who started the fashion because he was too fat to keep himself buttoned. In *The Road to Wigan Pier*, George Orwell made fun of one Socialist, an old Etonian: "He would be ready to die on the barricades, in theory anyway, but you notice that he still leaves his bottom waistcoat button undone."

Imitating the rich in financial matters makes more sense but can also be hazardous. In the 1820s, small stock traders came to regard Nathan Rothschild as the master of the London Exchange. "This meant that overt Rothschild sales or purchases could trigger a general flight from or into a particular stock," writes Niall Ferguson. So Nathan and his brothers employed multiple stockbrokers. They used one broker to make a show of selling a stock, producing a ripple effect among their imitators. Then they used another broker to quietly buy up the stock after their hapless imitators had driven down the price.

You Are What You Ape?

I do not really intend any of this to suggest that subordinates are stupid or easy victims for exploitation by the rich. The dynamic of the relationship isn't nearly so one-sided. Subordinates use the rich, as well as being used by them, and they are often merely positioning themselves for better things. The rich clearly know it, and they work to assert their control from the start.

The subtle fear of competition colors many job descriptions: A top Washington lobbyist, for instance, seeks a female personal assistant to help arrange entertainment for senators and heads of state. She must be "a smooth operator . . . somebody who can totally be trusted with everything," though the possibilities of the job are not necessarily the ones that spring to mind. "Nobody who has political ambitions themselves," the client sternly warns. "No one who wants to be out lobbying or partying with the principals themselves."

"You can't put a clean-cut guy with a rock star," says Janine Rush of Sterling Domestics. "They ask for somebody they won't be embarrassed to smoke pot in front of. People say, 'My wife won't let me have anyone too pretty, but don't send me somebody I'm going to be embarrassed by.'

Or 'My husband won't let me have a male in the house.'" A new assistant almost never enters a wealthy home without an enthusiastic statement of trustworthiness from some former wealthy employer. Most are also obliged to sign elaborate confidentiality agreements. "In this business, references are everything," says Rush. "If you don't have a reference that says you're discreet, that you've done a good job, it's almost impossible to jump onto the next level."

The subtle fears of the rich about their own staff may be entirely warranted. The relationship is often much more equal than one might suppose. The subordinates are the ones with common sense, the ones who manage the schedule and make sure things get done. It's easy for them to assume, on some level, that they're really in charge. This is particularly true with second- or third-generation rich, who sometimes act, in the words of one fund-raiser, as if they are "47 going on 17." But it can be true of the self-made rich, too, because of their acquired incompetence. So when subordinates jokingly refer to their employers not as their fathers but as their children, they may mean it.

No matter how orderly it may seem, every hierarchy seethes with envy, deception, status rivalry, and tension over who ultimately ought to be in charge. Primatologist Christopher Boehm parses submission by subordinate apes into at least three separate elements: In the beginning, there's fear and mistrust on both sides. Gestures of appeasement by the subordinate may lead gradually to a mutually useful long-term relationship involving protection against predators, help against political rivals, sharing of food, and routine healthy social contact. But throughout, the subordinate nurses an underlying desire to be the one in charge. Boehm writes: "The primatological literature is replete with accounts of ambitious subordinates who bide their time until the situation is ripe for rank reversal or a takeover of the top position." Even when the subordinate bows down in front of the alpha, he is often casting sidelong glances in the direction of opportunity. Because opportunity is everywhere, the lives of the rich are littered with faithful servants who steal precious objects, sell family secrets, embezzle, blackmail, cuckold, murder, and betray. In truth, the rich are sometimes acutely aware that their own family fortune got its start in just this entrepreneurial fashion. In the eighteenth century, the Foxes somehow rose from footman in one generation to Earl of Ilchester in the next. Likewise, the British earldom of Winchilsea and Nottingham was created through bribery; the family motto to this day is the rather defensive *Nil conscire sibi*, or "Conscious of no evil."

In our time, Barbara "Basia" Piasecka was a thirty-one-year-old immigrant of modest means when she became a chambermaid in the household of J. Seward Johnson, who was then seventy-three. She committed no crime, but with the help of her ample sexual appeal she displaced his previous wife and positioned herself to inherit $350 million. In turn, Nina Zagat, now best known as cofounder of the *Zagat Restaurant Survey*, found her way to a fortune as Basia's friend and lawyer, her "enforcer, her apologist, her henchman, her heavy." In turn again, Lech Walesa, the Polish labor leader, also latched onto Basia Johnson's hem. "I like Pani Basia as a woman, not only for her money," he enthused in *The New York Times*. "I like to kiss her and I am quite happy that she still lets me. And I'm quite interested to see how long she lets me." Not long, as it happened. No doubt Basia knew there would be plenty of other subordinates to kiss and cosset her, much as she had once kissed and cossetted J. Seward Johnson—a long, long line of them wheedling favors from her and looking sidelong for opportunity down to the final hours of her life.

7

Why Do Rich People Take Such Risks?

Display by Grandstanding

We are all worms. But I do believe that I am a glowworm.

—WINSTON CHURCHILL

TWO-THIRDS OF THE WAY INTO HIS AUGUST 1998 ATTEMPT TO FLY round the world by balloon, Steve Fossett ran into a thunderstorm at 29,000 feet above the Coral Sea and began to plunge uncontrollably through the darkness. Wind whipped his ruptured balloon from side to side. Hail hammered down. At 4,000 feet, Fossett climbed through the hatch atop his capsule and cut away fuel and oxygen tanks to slow the balloon's fall. Then he lay down on a bench to distribute the impact across his back. "I'm going to die," he said out loud.

I'd met Fossett the year before and he was mild and Midwestern, an Eagle Scout, a former computer systems engineer who'd built up a successful Chicago commodities brokerage, a multimillionaire with no particular need for publicity. So what was he doing falling out of the sky in a broken balloon? Why did Dennis Tito pay $20 million to get himself shot into space on a Russian rocket? What was billionaire Larry Ellison doing in the 1998 ocean race from Sydney, Australia, to Hobart, Tasmania, in which six sailors died amid forty-foot seas and ninety-mile-an-hour winds? Why, that is, do rich people do such dumb, dangerous stuff?

All of them, one way or another, were showing off. To put it in the biological context, they were engaging in display behavior. Animals do it all the time, and their displays, like ours, fall loosely into two categories: They show off with fine feathers, and they show off with risky behavior. Often, they do both at the same time. The broad-tailed hummingbird, for instance, is one of the oldest and flashiest seasonal residents of the Aspen area. The male of the species has a bright red throat-patch and metallic green feathers. But he also tries to impress rivals and potential mates by shooting sixty feet straight up in the air and back down again in a gaudy power dive, which he may repeat forty-five times an hour, his wingtips giving off a metallic trill urgent as a kid's bicycle bell.

But rich people aren't hummingbirds, and they tend to deny that they are engaged in display behavior at all, much as they also disavow an interest in money or power. They generally prefer to characterize their bold deeds and precious possessions as the happy outcome of personal passion and connoisseurship. Any suggestion that they might also be motivated by the urge to impress is merely callow and invidious. A dealer for Bill Gates and other Microsoft executives put it inelegantly: "They collect exactly what they want, and they don't give a shit about impressing anyone." A Dutch entrepreneur introducing a new custom sports car, the $157,000 Spyder C8 Spyker, also caught this frame of mind neatly: "We're building a car for people who have nothing more to prove, who are treating themselves to something beautiful."

Does any such person exist? If so, why is he driving around in a vehicle that pretends to be both a "spyder" and a "spyker," something undoubtedly fierce and predatory? A cynic would say that the only people with nothing to prove are already dead, and even death does not stop them from trying, to judge by the monuments and foundations they sometimes leave behind. Nor should they stop. The urge to impress is at least as powerful in humans as in other animals. It is *so* powerful we sometimes seem to be conducting an eerie dialogue, the living and the dead displaying for one another across the abyss. When the rich say that they don't need to impress anybody, they usually mean only that they have drastically narrowed down the list of people they are interested in impressing. Even when they think they have narrowed it down only to themselves, they are often still proving things to the ghosts of unloving fathers or of old teachers who thought they'd never make it. In this urge to display for dead mentors, the rich are a bit like the pronghorn deer, which has evolved to sprint at more than sixty miles an hour. No potential predator now living in the American West comes even close to that speed. But a species of cheetah used to live there, and some scientists believe the pronghorn is still running from the ghost of a predator that went extinct ten thousand years ago.

A critic of the idea of display behavior in humans could perhaps argue that animals have better equipment for the purpose than we do. The writer and naturalist Gerald Durrell was once taking Princess Anne on a tour of his zoo on Jersey in the Channel Islands, when they encountered a mandrill in full sexual display, its bottom like "a newly painted and violently patriotic lavatory seat," all blue on the outer rim and "virulent sunset scarlet" within.

"Wonderful animal, ma'am," Durrell said to the Princess, and then added, "Wouldn't you like to have a behind like that?"

Not even British royalty are so happily endowed. The human repertoire of biological display devices is limited and, at best, not terribly ornate: The size and proportion of different body parts, the symmetry of the face, the condition of skin and hair. Our repertoire of cultural display devices, on the other hand, is immense and compares favorably even with a mandrill's patriotic bottom. The rich can tell one another precisely who they are and where they stand by means of fashionable clothing, hand-stitched shoes, jewelry, fine foods and wines, great art, good music, fast cars, big yachts, private jets, thoroughbred horses, splendid homes, and beautiful friends. (Sir Philip Sassoon, a British millionaire who liked glamorous company, once got a letter from a friend: "It is Easter. Christ is risen. Why not invite him to lunch?") In fact, humans have an almost infinite variety of such displays, and unlike the relatively fixed display patterns of the natural world, our displays are utterly changeable. What is bold and fashionable one moment is foredoomed to become antiquarian the next. So it can easily seem as if our display behavior is entirely cultural and has nothing whatever to do with the natural world—-except maybe that we sometimes borrow our fur and feathers from animals.

Yet even this sort of secondhand display behavior has precedents in nature. For instance, the male bowerbirds of New Guinea are relatively drab, so they build great houselike bowers to entice potential mates. They also decorate their bowers with colorful *objets d'art.* Prominent among these *objets* are the ornate crests, fans, plumes, and tail streamers of a spectacular neighbor, the bird of paradise. To ensure that a visiting female understands how handsome this makes him, a male bowerbird will some-times pick up one of his decorations and hold it in his beak while pranc-ing, fluffing up his feathers, and flapping his wings to the beat of his own call. Wealthy women used to behave to much the same effect while wear-ing bird of paradise feathers in their Easter bonnets.

There is in truth no need to be coy about our display behavior. The urge to impress, to show off, to turn heads, to prove oneself over and over, is entirely natural. It is also entirely healthy. Except, of course, when it is fatal.

Risky Business

Fossett's *Solo Spirit* crashed into the Coral Sea hard enough for him to black out momentarily. When he came to, his capsule was upside down in the ocean, half-full of water and with flaming propane starting to torch through the hull. He grabbed his emergency satellite rescue beacon, scrambled out into his rubber raft, and then waited twenty-three hours to be rescued.

Fossett clearly was not one of those people who have nothing more to prove. Nor was he treating himself to something beautiful. The *Solo Spirit* capsule did not even have a window for watching the world drift by below, though Fossett could of course enjoy the scenery al fresco, atop his capsule, so long as he could stand the cold and the lack of oxygen, which were marginally worse than the ice-in-marrow interior. Everything about his round-the-world ballooning ambition was uncomfortable and dan-gerous, which goes against our ingrained notion of what being rich is all about: If you have the money, after all, why not just stay home by the fire at the vacation place in Vail? Why not become a patron, hiring people to balloon around the world on your behalf, as the Breitling watch company does? We act as if it is almost unnatural for the rich to accomplish things for themselves, as in the Hilaire Belloc poem: "Lord Finchley tried to change the electric light, / It struck him dead, and serve him right." Even to Fossett himself, drifting in his raft in the middle of a vast empty sea, his risky undertaking must have seemed like the inexplicable, even unnatural, quirk of one eccentric rich person.

But a taste for grandstanding is common in the natural world, too. For instance, antelope pursued by hungry cheetahs often leap acrobatically straight into the air, a practice called "stotting," when common sense says they should be sprinting for the far horizon. Even lowly guppies dance right under a predator's nose before darting away. So why do rich people and animals alike do such dumb stuff—stuff that is unnecessary, flamboyant, and often downright deadly?

The Handicap Principle

In search of answers, I found myself, at five o'clock one morning, rattling across an Israeli desert in a dusty little Peugeot with Amotz Zahavi, the elderly bête noire of the biological world. "This is a minefield," Zahavi said, indicating a fenced-off area just to our left. Then he veered right, both hands on the wheel, down into a wadi, or dry river bed. "So we won't go there." Zahavi is an ornithologist who has spent much of his life on the border between Israel and Jordan studying Arabian babblers, the philanthropic birds described in an earlier chapter. He is best known as the author of a far-ranging and highly controversial theory, the handicap principle, which attempts to explain, among many other things, why antelope stot, why peacocks carry splendid but cumbersome tails twice the length of their bodies, and why rich people engage in such extraordinary display behavior.

Zahavi's handicap principle holds that animals and humans alike prosper not *in spite* of our riskiest and most extravagant behaviors, but *because* of them. These behaviors are the way we advertise how prosperous, how fit, how fearless we are. And because the world is a jaded, cynical place, we have to incorporate a significant cost, or handicap, in our advertising to make it persuasive. Thus antelopes really are indulging in a dangerous waste of energy when they stot in front of a cheetah, but their willingness to risk it is how they tell the cheetah, Don't even bother trying. Among the rich, the handicap may sometimes involve risking one's neck, like Fossett in his balloon. Or it may entail some recklessly extravagant gesture, like the Australian Kerry Packer, in the summer of 2000, losing $20 million in a weekend playing baccarat in Las Vegas. Or like the industrialist William I. Koch, at an auction in another Nevada casino in the summer of 2001, almost doubling, to $2.4 million, the record price for a painting by the nineteenth-century cowpoke artist Charles M. Russell. Carrying off such gestures without a flinch sends a message to the world: You can't touch me.

When Zahavi first proposed his handicap principle in the 1970s, the biological establishment reacted as if something bad had gotten stuck down its throat. In the first edition of his book *The Selfish Gene*, the Oxford evolutionist Richard Dawkins called the handicap principle "maddeningly contrary" and he wrote, with a cadence and clarity rarely seen in scientific writing, "I do not believe this theory." Robert Trivers, the Rutgers University evolutionist, kidded Zahavi that if you took his idea to the logical extreme, you'd end up with a bird in which both sexes fly upside down to show each other how good they'd be if they were flying right side up.

It didn't help that Zahavi, a prominent conservationist who'd switched to biology in midlife, was himself maddeningly contrary. The standard scientific practice of testing ideas with mathematical models was alien to him. He hatched his ideas based on observation and intuition alone, and he questioned the intelligence of those who failed to embrace his conclusions, often including prominent scientists who had risen by more traditional routes. "But they *do* fly upside down," Zahavi said, when I reminded him of Trivers' joke one day, and he proceeded to rattle off bird species in which somersaults and other inversions are part of the mating display. Indeed, Zahavi's own conversation often seemed like a prime instance of the handicap principle, a contra dance at the edge of the unlikely. He summed up his big idea in a phrase: "Something can be good because it's bad."

In spite of Zahavi himself, evidence for his handicap principle began to accumulate. One study showed that African wild dogs and hyenas in fact avoid animals that stot, apparently because nonstotters are easier to catch. More important, an Oxford biologist named Alan Grafen demonstrated with a mathematical model that the handicap principle made sense, in evolutionary terms. In the second edition of his book, Dawkins grumbled about the possibility that "theories of almost limitless craziness can no longer be ruled out on commonsense grounds." But he added, "If Grafen is right—and I think he is . . . it might even necessitate a radical change in our entire outlook on the evolution of behaviour."

Sexual Selection

In fact, Zahavi's handicap principle addressed one of the central problems in evolutionary thinking. Charles Darwin is of course best known for his theory of evolution by natural selection. But in his 1871 book *The Descent of Man*, he proposed an equally important idea, largely ignored until mid–twentieth century (perhaps because it challenged male supremacy, an orthodoxy even more sacred than the biblical account of creation). Evolution by *sexual* selection holds that genetic change is influenced at least as much by the ability to attract members of the opposite sex. The trouble with these two ideas is that they often seem to contradict each other.

Natural selection essentially means that nature weeds out unfavorable traits by killing individuals who display them. Thus an arctic fox with a patch of red fur spelling the words "Eat Me" on his side would quickly end up in a polar bear's belly. At the same time, individuals with more favorable traits, like plain white fur for camouflage in snow, tend to survive and reproduce.

But in the vast majority of species, females control mating and do most of the sexual selection, and they often have an irresistible attraction to the male with the "Eat Me" sign on his side. That is, they seem to select males for traits that make them less likely to survive. In peacocks, for instance, the upkeep on a big flashy tail requires the male to waste huge amounts of energy. It also inhibits his ability to fly and makes him more vulnerable to predators. The female's own drab coloration attests to her abiding faith in the value of camouflage. But she will nonetheless choose a male with a bigger, showier tail almost every time. The natural world is in fact full of females falling hard for such seemingly stupid male display behavior, including bright feathers, big antlers, and bombastic courtship rituals. The same is of course true in the world of the rich. When the bald, cigar-wielding billionaire Ron Perelman first took an interest in Patricia Duff, for instance, he phoned her from his private jet at Los Angeles International. He didn't merely ask for a date, but told her the engines were running and would continue to run until she joined him. And Duff, quivering before such profligate displays, ultimately said yes, oh, yes.

To be fair, male sexual selection sometimes forces irrational displays on females, too, and female displays also make males go all aquiver. For instance, small breasts are just as good as big ones for nursing babies, and in our ancestral environment flat-chested females would also have been better equipped for climbing trees and running from predators. So the logic of natural selection argues that we ought to be a double-A-cup species. But men like larger breasts. Having them is of course a handicap, a cost ancestral women undertook, according to Zahavi, to display greater nutritional fitness in the form of visible body fat.

Given the ruthless efficiency of the natural world, how could such costly displays have evolved in the first place? The classic explanation enshrined in biology textbooks is the "runaway process," a variant on sexual selection theory devised by the British mathematician R. A. Fisher. Let's say ancestral peacocks started out dull-eyed and unbushy-tailed. Then, through some minor genetic shift, a few females developed a hankering for males with slightly longer tails. If this flourish happened to occur in bigger, better males, then females who chose them would probably rear more offspring. Longer tails would proliferate in males, and the preference for them in females. So far so good. But the conventional understanding of Fisher is that females soon come to focus single-mindedly on this trait, regardless of a male's overall quality. Thus begins the cycle of runaway one-upmanship: Over thousands of generations, the peacock's tail continually lengthens and spreads out until no self-respecting male can get a date without a huge fan rattling on his rear end. Natural selection steps back in to stop the runaway process at the point where the quest for better sexual ornamentation is attracting enough predators to kill hapless males.

The trouble with this idea, a student of Zahavi's argued one day, is that sexual selection isn't that whimsical. In the real world, females rigorously test males to see if they will make good mates, and males of course test females, too, though perhaps less rigorously. Zahavi came to think females would choose an exaggerated trait only so long as it reliably signals that the male would make a better mate. So what good does a huge tail do for a peacock? Why would Irish elk have developed antlers with a twelve-foot span? And why is Steve Fossett always banging around the planet in his balloon?

Zahavi came up with a theoretical answer, then confirmed it one day while watching his babblers. He was trying to figure out why the babblers kept shouting at a hawk, instead of just quietly hiding in a bush. "And then I realized, *they are talking to the predator.*" They were taking on a handicap by revealing their presence to the hawk—and they were advertising that a surprise ambush wasn't going to work. It convinced Zahavi that flamboyant handicaps—the babblers' shouting, the peacock's tail, the elk's antlers—actually serve a useful purpose: The same displays that attract females also discourage predators and rival males.

This was the basic idea Oxford biologist Alan Grafen refined into the language of theoretical mathematics. He started with the assumption that females could use handicaps as a gauge for selecting the fittest males. For example, the arctic fox with the "Eat Me" sign wouldn't survive a single winter unless he also happened to be incredibly quick or cunning. Then Grafen calculated the advantage to the daughters in inheriting greater fitness, without expressing Daddy's dumb male handicap displays. Finally, he added in the advantage to the sons who get saddled with the displays but thus attract more females and scare off more rivals. Grafen fine-tuned his model using an idea from Zahavi that other scientists had overlooked: Sons actually express Dad's handicap displays only in proportion to their own physical abilities. Thus Dad may stot like a champion. But if Number Two Son lacks zip, he soon drops out of the evolutionary equation in the form of fast food. When he put all these calculations together, Grafen's model showed that, over time, the handicap principle should result in more offspring. The instinct for risky display behavior, and the tendency to be impressed by it, should thus proliferate within a species.

Panache

The idea that Fossett was showing off seemed wildly improbable to me, at first. He was a middle-aged businessman with thinning gray hair, thickish waist, and a Boy Scout manner; in a crowd you would pick him out as a bystander. At one point when weather conditions were right and several balloonists were attempting to launch at about the same time, his competitor, that creature of limelight Richard Branson, showed up for a Fossett launch at Busch Stadium in St. Louis. Branson's own flight had aborted a few days earlier in North Africa. He'd had time to skip home to England for a press conference, then down to Switzerland for another press conference before the launch of a Breitling balloon, and now Branson came sweeping into Busch Stadium beaming with the glorious camaraderie of it all. A cameraman stayed close to record the swashbuckling manner, the swept-back golden mane, the big grin parting his wolf-man beard. Branson delivered a sound bite about the incredible bravery of the pilot about to fly off in this tiny capsule. Then a nondescript man standing beside him said, "Well, we're ready to go."

"Oh, well," said Branson, "I was just hoping to have a word with Steve."

"I *am* Steve," the other man said.

For Branson to be a showoff was given. But Fossett? Yet any doubts I had on this score vanished the day Fossett told me how he'd gotten started in ballooning: It happened during a visit with his wife to Paris. He was looking at a $250 scarf from Hermès, depicting some of the great pioneers of aviation. It occurred to him that ballooning was by far the oldest form of human aviation, yet no one had ever succeeded in taking a balloon around the world. Attempting to become the first was a bid for status made even riskier, and thus more glorious, by Fossett's initial ignorance of ballooning. Up to that point, he had never even ridden in a balloon as a passenger. The ambition was so outlandish that he could not even bring himself to inform the ground crew on his first long-distance flight that he was actually intending to continue on around the globe. But he wanted his face on a scarf like that.

Fossett's ballooning was of course a prime instance of the handicap principle at work. It was a display of prowess, like innumerable other risky adventures by the rich. In Darwinian terms, it was a sexual display, and the very unlikeliness of Fossett as its subject suggests just how pervasive such displays really are among the rich. (Display behavior is of course pervasive in the rest of human society, too. But only the rich, by and large, can pick up the tab for a round-the-world balloon, or other flamboyant displays.)

The topic of sexual display also suggests that several strong caveats are in order: First, because much of the extravagant showing off in the natural world gets done by males trying to woo choosy females, it's easy to get the mistaken impression that risky display is largely a male phenomenon. In our species, females are still the choosier gender. But males, particularly if they are rich or famous, can also be choosy. So both genders indulge in sexual display. A woman may be as likely as a man to buy a Van Gogh or to trade on the status of owning a mid-eighteenth-century slant-top Chippendale desk, say, or a $4,000 Lana Marks handbag ("Miss Marks works only in lizard, alligator, crocodile, and ostrich. The finest materials," a sales clerk advised me. "All the hardware is brass dipped in 24 karat gold. Repeatedly. That's how it gets the deep yellow color. This is the Princess Diana bag, commissioned by her before her death.") Women also indulge in risky displays, as when Susan Butcher won the Iditarod dogsled race and beat the (male) speed record by thirty-one hours. The socialite and publicity-seeker Sandy Pittman was of course also engaging in Zahavian display behavior when she attempted to become the first woman to climb the seven highest summits in the world, though she distributed the handicaps rather too unevenly by having her sherpas on Everest carry her eighty-pound satellite phone, her espresso maker, and her "ample supply of Dean & DeLuca's Near East blend," before one of them actually short-roped her to the summit. Women also engage in risky displays on more traditional lines, as when actress Jennifer Lopez showed up at the Oscars in a dress that defied gravity, anatomy, and common sense. We now know that she held it up with glue. Why?

Because display behavior really impresses only when it dances on a dangerous edge. Or as the rich men and women who race ocean sailing yachts put it: If something doesn't break, you built it too strong.

The second, and larger, caveat is that hardly anyone buys a Chippendale desk, or any other display item with the possible exception of breast implants, under the impression that this will lead in a direct arithmetical progression to orgiastic sex. Even if the scientific term is "sexual display," most of our display behavior is not consciously sexual. The urge to make extravagant displays certainly evolved through the give-and-take of sexual selection, and it also influences present-day sexual choice, which is why scientists use the term. But display has become an end in itself, even for people who are not actively out in the sexual marketplace. For that matter, display is often an end in itself among animals, too. This was never more evident to me than one day, during my visit with Zahavi, when I saw a bird called the houbara bustard in a cage at the Tel Aviv University research zoo. He was about the size of a rooster, but showier, with a pile of white feathers like an ascot at the base of his throat, black feathers fanning out from either side of the neck, and a big white headdress. Suddenly, the bird threw his head back between his wings, causing the feathers of the ascot to open in a huge fan where his head used to be. Then, completely blinded by this magnificent display, he began strutting like a drum major. When he reached the end of his run, he flipped his head up, flattening the neck plumage and bringing his beak together with a clap. Then he turned around, flipped his head back, and did it again, over and over. One can hardly imagine what kind of show he would have put on had there actually been a female in sight. Display behavior was simply something he did, regardless. He had this look in his eye as if to say, Don't I look *fine?*

We also like to look good, to dress well, to live in attractive homes, for their own sake, even when no one is watching. We do it, or so we like to think, to please ourselves. If this happens to please other people, too, and make us more attractive to them, that may seem more or less incidental. Even if we somehow manage to be blissfully unaware that beautiful displays often lead to congenial relationships, other people, particularly members of the opposite sex, take notice. Like social dominance and serotonin levels, the things that please us often unconsciously reflect the good opinion of the outside world.

Fossett, when I caught up with him by E-mail, had been rescued from the Coral Sea, was just back from lopping seventeen hours off the New York-to-Miami speed record in sailing, and was about to head out the door to crash-land yet another balloon, this time just over the Andes. (Having lost the race to complete the first round-the-world balloon flight, he was still trying to become the first to do it solo, and in mid-2002, he accomplished this goal.) He was, as always, boyish and polite. I asked him why he did it—the ballooning, the ocean racing, a run in the Iditarod, the three attempts to summit Everest, and so on and so on and so on. His answer was unenlightening: "I've never gotten tired of logging another achievement." I didn't ask him whether he was doing it to impress his wife, his business rivals, or the ghost of some distant mentor. I don't think he really knew what made him tick, any more than the rest of us do. Like the hummingbird doing power dives, or the houbara bustard blindly strutting back and forth in his cage, he was simply caught up in the mindless, unending, and frequently splendid animal drive to prove and display one's worth.

8

Inconspicuous Consumption

Display by Muted Extravagance

Fine feathers, they say, make fine birds.

—Isaac Bickerstaffe *(1735–1812), Irish dramatist*

ONE DAY IN MONACO I STRUCK UP A FRIENDSHIP WITH AN ELDERLY countess, who wore a diamond the size of a strawberry on her finger, three long strands of pearls, a gold bracelet of elephants joined trunk-to-tail, diamond earrings, a brooch, and a large gold pin holding her broad-brimmed hat in place. She was smoking a Cartier cigarette from a gold case, and her eyes brimmed with the jade color of the sea. Her first husband collected Impressionists, she told me, and she still had a Picasso, Blue Period, in a vault somewhere. But her own tastes were simple: "I like jewels." She was staying at the Hôtel de Paris for the refractory possibilities of the light flooding in through the lobby windows. "I want them to see my jewels," she said. She was as happy as a child. "I like it to gleam."

This was exactly how it was meant to be in Monaco, where everything that glitters is, if not gold, probably diamonds. It was also conspicuous consumption in its most straightforward form. Thorstein Veblen, who coined the term in his 1899 book, *The Theory of the Leisure Class,* regarded conspicuous consumption as the key to social status for the rich: "The basis on which good repute in any highly organized industrial community ultimately rests is pecuniary strength; and the means of showing pecuniary strength, and so of gaining or retaining a good name, are leisure and a conspicuous consumption of goods." Veblen's phrase has entered the language and come to serve as a buzzword for the most splendid and absurd excesses of the rich: For instance, the dinner put on for a visiting dignitary in 1550 by the Italian merchant Gaspare Ducci, at which he served oysters with their shells plated in gold. Or the country manor built in 1890 by the British swindler Whitaker Wright that included an underwater billiards room. Or the 1976 flight Elvis Presley took in his private jet from Memphis to Denver and back, consuming 5,500 gallons of fuel, to purchase and eat a sandwich made from a loaf of bread slit down the middle and slathered with peanut butter, jelly, and a pound of crisp bacon.

Conspicuous consumption has become such a part of our language that people often mistakenly regard it as the only form of display behavior among the rich. Partly because Veblen was a social satirist as much as an economist, an absence, or even outright defiance, of commonsense has come to seem like a defining characteristic of conspicuous consumption. It's about the self-indulgent rich at their most unnatural—for instance, the Sultan of Brunei giving a fiftieth birthday party for himself, featuring that *ne plus ultra* of wretched excess, three separate concerts by Michael Jackson.

Among the least commonsensical aspects of conspicuous consumption is the willingness of the rich to pay more when comparable merchandise is available for much less. For the leisure class, Veblen argued, price tag is essential for status. "A beautiful article which is not expensive is accounted not beautiful," Veblen wrote. A mass-produced spoon may have a graceful design, he suggested, and a hand-crafted silver one may be clumsy and ornate, but we almost invariably treat the silver spoon as the heirloom. Economists now refer to this idea as the Veblen Effect. It suggests that the rich prefer a high price because it advertises that they can afford, say, the canvas high-top sneakers by Gucci, with chinoiserie needlepoint, while simultaneously excluding those who can't.

The Veblen Effect isn't just an academic idea. It actually shapes the way companies price the merchandise they put on the shelf and it suggests that, contrary to expectations, cutting prices for high-end items may actually *decrease* sales. The Four Seasons Hotel group, for example, has sometimes shut down a hotel in a slow market, rather than cheapen the brand by discounting room rates. During the slump in tourism after the September 11, 2001, terrorist attacks, many high-end hotels enticed guests by offering to add an extra night to their stay at no cost, rather than risking the psychological downgrade from actually cutting rates. Pierre Cardin, in apparent ignorance of Veblen, went from haute couture in the 1960s to low-cost and no prestige in the 1980s.

If the behavior of the rich is sometimes counterintuitive, it is not, however, in the least unnatural. Veblen's conspicuous consumption and Zahavi's handicap principle are, in fact, variations on the same idea: the one arrived at in the context of human luxury; the other, independently, in the context of animal behavior. Veblen the economist was apparently ignorant about the natural world. His ideas about evolution were peculiar, often dwelling on leisure-class ethnic types like the "dolichocephalic-blond." Zahavi, the biologist, in turn, was unaware of Veblen until after he had published the handicap principle. This ignorance was entirely pardonable on both sides: Who would've thought that rich people and wild animals behave the same way? But if the one theory is principally about possessions, and the other about biology and behavior, both argue that the essential quality in display behavior is extravagance, or what Veblen once called "conspicuous wastefulness."

Waste Is Beautiful

The upshot is that wasteful display now appears to be entirely natural. This promises, at least at first glance, to be good news indeed for rich people. According to some followers of Veblen and Zahavi, biodiversity itself derives from the proliferation in the animal world of wasteful sexual displays, like the flashing light of the firefly or the brilliant color shows put on by courting squid, which bend and turn together like Gene Kelly matching step for step with Cyd Charisse. This is the giddier side of evolutionary psychology, having to do with what one researcher terms the "wonderland of wasteful sexual signaling."

"Many evolutionists now recognize that almost everything in nature which we find beautiful or impressive has been shaped for wasteful display, not for pragmatic efficiency," writes Geoffrey Miller, of University College, London. "Flowers, fruits, butterfly wings, bowerbird nests, nightingale songs, mandrill faces, elephant tusks . . . and human language have all been shaped by sexual selection." In nature, sheer profligate wastefulness is often an essential part of the display. It's not enough, for instance, for the broad-tailed hummingbird to have colors better than a Tiffany brooch; he really only impresses if he also takes on the conspicuous cost—the handicap, in Zahavi's terminology—of his forty-odd power dives an hour. Making this wasteful display requires him to remain svelte and agile, so the hummingbird actually starves himself all day long, through power dive after power dive. Then, like a rich woman's walker letting out his cummerbund at the end of the night, he goes on a twenty-minute binge, cramming down nectar until his belly sags and his weight increases by a third with the energy he needs to survive for another day. A male who cannot keep up this extravagant lifestyle gets no females.

Miller, an exponent of the handicap principle, argues that the same urge to wasteful display is behind our taste for shiny cars, designer dresses, large houses, or a Van Gogh on the wall. The sheer wastefulness of these possessions is precisely the point. If we could have the Van Gogh for the asking, Miller suggests, many people would no longer want it. They buy such paintings to let people know—tastefully, of course—that they can afford to have $70 million hanging on the dining room wall. Likewise, a Ferrari 360 Spider is a status symbol mainly because it costs $140,000. Drop a zero (and change the name to, say, the Little Miss Muffet) and it's just a noisy, uncomfortable sports car.

Forgive me, I exaggerate. The Van Gogh is undoubtedly fine art, and the Ferrari can go from zero to sixty in a nanosecond; each has its own intrinsic value. But how much value, relative to price? Miller compares Sennheiser Orpheus Set stereo headphones, which cost $15,000, with Vivanco SR250s, which sound almost as good, but cost about $40. "The marketing people at Sennheiser know," he writes, "that Orpheus Sets are bought mainly by rich men, young or middle-aged, who are on the mating market, openly or tacitly. Their 400-times greater cost than the Vivancos is a courtship premium." That is, the cost is the handicap. "While the Vivancos are merely good headphones, the Sennheisers are peacock's tails and nightingale's songs . . . We want Sennheisers not for the sounds they make in our heads, but for the impressions they make in the heads of others."

Miller goes on to argue that the wasteful displays of the rich may actually be beneficial: Is wastefulness "the best attitude for fostering sustainable human societies?" he asks. "Surprisingly, biology suggests that the answer may be yes." One advantage of *our* wasteful displays, Miller argues, is that they are more durable. The peacock's fan dies with the peacock, but the Van Gogh is likely to serve as a sexual display for generations. In Miller's view—which can sound at times like an odd blend of Reaganism, Darwinism, and environmentalism—conspicuous consumption by the rich actually leads to trickle-down display: "Although a physical product can only be owned by one individual, its sexual status value can be enjoyed by anyone associated with its production, financing, marketing or consumption." Owning the Van Gogh begins to seem downright egalitarian. It becomes a bragging point even for the guy whose job is to hang it on the dining room wall.

Miller believes wasteful sexual display may also prove environmentally beneficial by distracting people from sex itself, or at least from reproduction. "The more time and energy you waste on showing how sexy you are," he writes, "the less time and energy you have for raising offspring. . . . Instead of spending our 20s taking care of our first six toddlers, as our ancestors would have done, we spend our 20s acquiring a university education, launching our careers, buying stuff, going to movies, taking vacations and worrying about status. Conspicuous consumption or conspicuous children, it is difficult to attend to both."

Well, OK, there's also some bad news here for rich people. Miller's argument to the contrary, many of our wasteful sexual displays are not nearly so durable nor so environmentally innocuous as the Van Gogh or the Sennheiser headphones. Nor are they necessarily so ineffectual at the raw Darwinian business of genetic proliferation. Wilt Chamberlain was engaging in wasteful display when he had the bedspread and conversation pit in his Los Angeles mansion covered with the muzzles of several thousand arctic wolves. It would probably take a certified public account to track the wolf-to-woman ratio during his residence there, but Chamberlain himself once estimated that over the course of his life he'd had sex with twenty thousand women. Apart from the wolf muzzles, construction of the basketball star's love nest on a hilltop in Bel Air also required five freight-car-loads of old-growth redwood. And here's the trickledown: When the estate recently went on the market, a Chamberlain pal billed it as having "the cachet of being a house where a famous man who had a lot of women lived." In the argot of the real estate agent, this spells T-E-A-R-D-O-W-N.

In the same gaudy spirit, a Hollywood producer recently impressed a few friends by flying them in his private jet down to the Galapagos, where he had a crew filming an environmental documentary. They hung out for a couple of hours, then got bored and flew home, at a total cost the hapless film crew estimated to be more than $100,000. Lunch consisted of Darwin's finches served on a bed of cracked wheat and celery in a mustard vinaigrette. (Oh, all right, that last bit is a lie.)

Pro Bono Showoffs

Yet Miller makes an important point: Contrary to everything our mothers taught us, showing off is frequently beneficial, not just for the individual making the display, but for the rest of us, too. J. Paul Getty, for instance, purchased art as a means of transforming himself from an oil man into an emperor. (Many rich people feel a special affinity with Louis XIV or Napoleon, but Getty told friends he was the reincarnation of Hadrian.) Having "the art they admired in your house," was, according to Getty's art adviser Federico Zeri, one way to attain equal stature with old money sorts. But vanity passes, and the enduring benefit is that common roughnecks can now visit the J. Paul Getty Museum in Los Angeles to enjoy his collection at no cost. John D. Rockefeller, Jr.'s, astonishing philanthropic endeavors were motivated by the same innate human drive to display; they represented a show of virtue by a son intent on dispelling his father's diabolical reputation. But the public benefited immeasurably from the creation of Grand Teton National Park, Acadia National Park,

Redwood National Park, Williamsburg, The Cloisters, The Museum of Modern Art, and many other national institutions.

For philanthropic display to provide a public benefit is unsurprising, though not necessarily axiomatic. But even flagrantly wasteful showing off, without the slightest pretense of serving the public good, can sometimes be beneficial in ways that are more substantial than hand-me-down sexual display. The familiar rationalization of the rich is that the economic benefit of the $500 dinner at Alain Ducasse trickles down to the busboy, the $1 million solitaire makes a Merry Christmas for the diamond miners in Sierra Leone, and the $45 million Gulfstream V brings joy to the guy who gets to pump in 6,000 gallons of aviation fuel during a Stage One air-quality episode at Burbank Airport. Or as the economist John Kenneth Galbraith once put it, in his unkind definition of the "trickle-down theory," if you feed the horse enough oats, "some will pass through to the road for the sparrows." And yet the sparrows surely feed.

Beyond immediate economic benefits, though, showing off by rich people also often leads to permanent improvements in the way the rest of us live. Brian Hayden, whose work on competitive feasting and the origins of agriculture figured in an earlier chapter, writes that the competitive display behavior of the rich has resulted in "the transformation of numerous prestige technologies into practical technologies." The rich, or the people Hayden terms Triple-A aggrandizers, were the first to introduce the use of textiles, metals, open ocean boats, leather shoes, finely milled grains and white bread, plastics, plumbing (America's first flush toilet was at the Reading Room, an exclusive men's club in Newport, Rhode Island), ceramics, glass, automobiles, writing, and books. It isn't that the rich themselves necessarily invented any of these new technologies; but they served as patrons for the artisans who did. By making these innovations part of their conspicuous consumption, they also induced emulation by the rest of society. Hayden credits the process by which the playthings of the rich get transformed into the bread-and-butter of ordinary people with producing "a progressive but dramatic increase in the standard of living over the past ten millennia."

Lest all this seem too industrious for a supposedly idle class, the display behavior of the rich also gave us the institution of the weekend. British prime minister Arthur Balfour, an aristocrat whose family owned

87,000 acres in Scotland, possessed wealth, good looks, great charm, and "the finest brain that has been applied to politics in our time." He also liked to play golf. In the 1880s, he arranged his schedule, and the nation's, to allow for two full days of it a week. "Such was his magnetism," writes the historian Barbara Tuchman, "that Society followed where he went, and so the custom of the country-house weekend was born." The idea of the weekend trickled down to the working class sometime after.

Finally, all this suggests a far more arguable benefit: That glimpses of splendor are somehow good for society as a whole, that magnificence at the top of the social hierarchy dignifies even those at the bottom. Our ethical and religious readings teach us the opposite, that it degrades us all for some people to grow fat and wasteful while others go hungry. Political history also argues that too flamboyant a display of wealth risks the guillotine. But bear with this argument for a moment: We are all afflicted, like other social primates, with a counting-house instinct for social status. We take secret comfort from meticulously calibrating the people and behaviors deemed below us on the social ladder, and we yearn blissfully toward distantly attainable wonders up above. There is hardly anything more dispiriting than the phrase, "It doesn't get any better than this." We like to think we are getting ahead in the world, which means we need an ahead to be getting to. This is true even in the most absurd contexts. When he staged fashionable parties in the 1990s, for instance, the British public relations shill Matthew Freud liked to arrange a series of progressively more exclusive VIP rooms, and the inner sanctum at the end was typically empty.

The display behaviors of the rich sometimes also deliver better things in ways that are not at all absurd. The splendid wastefulness of the rich has bequeathed us some of the most beautiful things on Earth—Botticelli's *Birth of Venus* commissioned on behalf of Lorenzo de' Medici, for instance, and Mozart's *Requiem Mass* commissioned anonymously by Count von Walsegg zu Stuppach, apparently so he could palm it off as his own work. So long as their wasteful displays aspire to such grandeur, showing off by the rich undoubtedly provides a service to humanity. Anything less, of course, and the rich themselves are acutely aware that they risk twisting on the treble hook of religion, ethics, and politics.

But whether their display behavior does any good for the rest of us is

ultimately irrelevant. All natural history cares about is whether it does
~ood for the individual making the display, and, if so, how.

Rites of Admission

One reason the rich pay such close attention to the signs and symbols of
wealth is to help them determine if a person is a member of their cultural
subspecies. Christopher Rockefeller, for instance, was not. He arrived at
the studio of artist Gines Serran-Pagan in Southampton, Long Island, in
the summer of 2000 claiming to be a French-born heir to the Rockefeller
fortune. He was driving a gold Mazda 626, which struck Serran-Pagan as
odd for a Rockefeller. But he quickly indicated his intention to buy a half-
dozen Serran-Pagan paintings, for a total of more than $500,000, the sort
of thing that encourages one to suspend judgment at least for the
moment. Over dinner, Rockefeller boasted about buying a $34 million
yacht and of being friends with noted playboys Bill Clinton and the late
Dodi Al-Fayed. Serran-Pagan thought he caught "the whiff of the street"
about his guest. Presumably it did not help that Rockefeller was muscular,
had an eagle tattooed on his arm, and spoke with a thick French accent.
Also, Rockefeller claimed that to transfer his payment, he needed Serran-
Pagan's bank account number; this was about as subtle an entrée for a
scam as those E-mails one is always getting from strangers in Nigeria: "In
order to transfer out $26 million from our bank, I have the courage to ask
your cooperation to handle this important business believing that you will
never let me down either now or in future." But what finally convinced
Serran-Pagan his guest was a fraud was his ignorance about wine, that
stumbling block for so many nouveaux riches. Serran-Pagan had poured a
cheap California jug wine into a ceramic pitcher and served it hoping no
one would notice. "Wonderful," Rockefeller enthused. "A Bordeaux, no?"
When Rockefeller was subsequently arrested for trying to beat a bed-and-
breakfast bill, he turned out to be a con man named Christophe Rocancourt, with no connection to the Rockefellers, wanted in multiple
jurisdictions on charges which currently include fraud, forgery, grand lar-
ceny, bail-jumping, sexual assault, and assault with a deadly weapon. The
moral of the story is, of course, know your wines. All else can be forgiven.

Knowing the right people, places, and pleasures—the sorts of things a rich person ought to know—serves as the only reliable badge of admission among the rich. This is especially so when billionaires dress in moth-eaten sweaters and drive twenty-year-old Jeeps. It may have been easier in the past. Sumptuary laws, from ancient Rome and dynastic China to colonial New England, attempted to regulate expensive displays according to class and income. In some places, peasants could not wear any colors except black or brown, rather like buyers of Henry Ford's Model T. Silk and linen were a mark of aristocracy, with silver buttons reserved for some ranks and gold for others. "Every costume was to some extent a uniform revealing the rank and condition of its wearer," according to one historian of such laws. In places, rank determined how many candles a person could carry in his lantern at night, and those with fewer candles were obliged to give way to literal social luminaries.

Such subtle distinctions, or badges of status, aren't just some human aberration. They are a way of life in the animal world. Among house sparrows, for instance, a large black bib is a mark of high status in males, and when females step out on their nest mates, they usually pick a paramour with a big bib. Great tits, a common bird species in the Old World, have a dark band on their chest. Males with a broader band receive the deference of males with a narrower one, and they are also more successful at defending territories and reproducing. Zahavi argues that these signals are not random but actually correspond to real strengths of the individuals who display them.

For a species as clever as ours, faking a badge of status might seem as if it ought to be easy. Assuming the identity of a Rockefeller was a huge success for Christophe Rocancourt, at least at first; he allegedly swindled several hundred thousand dollars from gullible Hamptonites wowed by the name. But fakers usually fail in the long run, as scientists have demonstrated in a series of somewhat bizarre forays into what might be termed the falsey factor. In one experiment, for instance, a researcher sewed artificial plastic swords onto the tails of male platyfish and found that females preferred males with long tails. The Danish ornithologist Anders P. Møller performed a similar experiment on European barn swallows. These beautiful songbirds have forked tails, and some are slightly longer than others. So Møller glued an extra few centimeters of feather on the tails of some males, and cut a few centimeters off the tails of other males. The males with longer tails found mates more easily, and they also got to indulge in more extramarital hanky-panky. Faking the badge of status translated directly into more sex.

But the Møller experiment, and a similar study by British researchers, also contained a surprising moral: The artificially enhanced tails impaired flight, and the fakers could no longer hunt insects as effectively. None of them returned from the winter migration the next spring. Moreover, DNA testing revealed that only about 50 percent of the offspring in the fakers' nests had actually been fathered by them. The females had some-how figured out that they'd been duped, maybe because the fakers simply did not bring enough food to the nest, and they went elsewhere in search of better genes. By contrast, males with naturally long tails fathered about 95 percent of the young in their nests. In other words, the badge of status only worked for individuals who were entitled to wear it.

Like barn swallows, humans routinely present one another with badges of their status. Even within their own ranks, the rich constantly scan for fakers, frauds, misfits, and people who merely overstate their status. During World War II, for instance, Robert Wood Johnson, Jr., left his job as president of Johnson & Johnson and served on the U.S. War Production Board, wangling the rank of brigadier general. For the rest of his life he insisted on being addressed as General Johnson, according to biographer Barbara Goldsmith; he also dropped "junior" from his name lest it detract from his air of social preeminence. In the same spirit, his sister Evangeline had married a Russian prince named Alexis Zalstem-Zalessky. "Uncle Bob would sit back," a niece recalled, "and say, 'Damned if I can understand why that woman insists on calling herself a princess, she's no more a princess than the man in the moon.' I remember saying to him, 'Well, I guess it's the same reason you call yourself a general.' "

The methods by which the rich test one another's status and fitness are by no means purely verbal. When Consuelo Vanderbilt, the Duchess of Marlborough, visited Russia in 1902, she attended a ball given for the Court of Czar Nicholas II at the Winter Palace: "With the first strains of a mazurka, the Grand Duke Michael, the Czar's younger brother . . . invited me to dance. It was a very different affair from the mazurkas I had learned at Mr. Dodsworth's class. 'Never mind,' he said, when I demurred, 'I'll do the steps,' and he proceeded to cavort around me until I was reminded of the courtship of birds."

Or, better still, the courtship of fruitflies. Certain strains of *Drosophila subobscura* suffer, like grand dukes and many other aristocrats, from inbreeding. One of the earliest studies of female mate choice found that female fruitflies rejected males that could not perform the normal courtship dance. The clumsy dancers were typically inbred males, which, even when they somehow managed to find a mate, produced fewer viable offspring. So using the fruitfly mazurka to screen out Drosophilan dorks and wankers helped the female improve her reproductive success. The elaborate ballroom dances popular during the height of the aristocracy apparently served the same function for human debutantes. Inbreeding was always an issue. On meeting the Czar, for instance, Vanderbilt was struck by "the extraordinary likeness that he bore to his cousin, the Prince of Wales, later George V. He had the same kindly smile, half hidden by a beard, the same gentle blue eyes and a great simplicity of speech and manner." Through intermarriage upon intermarriage, Nicholas and George had the same ancestors, and altogether too few of them. Hence perhaps the simplicity. Another grand duke, Constantine of St. Petersburg, testified unwittingly to the importance of skillful dancing as a display of genetic well-being. He was faced with an arranged marriage to a member of the Coburg family, which had intermarried so often and so successfully with European royalty that it was also deeply inbred. Carefully observing his bride, Constantine remarked: "If it must be so, I will marry the little ape. It dances very prettily."

Among ruling families of India, a ritual in which the groom's female relatives cleansed and anointed his prospective bride served as a far more explicit sort of biological screening: "The groom's family wants to have a good look at her whole body," a witness told oral historian Charles Allen, "to find out if there are any defects, no moles in the wrong place or concealed birthmarks. If that were found then someone would come running to say, 'Cancel the whole thing!' "

The presentation and testing of badges of status is of course considerably different in modern society. Titles and other traditional indicators of good breeding have suffered a considerable decline in value. Inbreeding is seldom an issue any more, and young people generally do not allow themselves to be inspected in the nude by their prospective in-laws (though visiting the family at the beach house accomplishes almost the same thing). So it is easy to get the impression that the only reliable badge of status left is money, or conspicuous consumption. According to this way of thinking, the single best way for a man to silence a prospective bride's skeptical family is to show them a thumping big pot of money on the other side of the marriage bed. (And don't let the bride see the prenuptial agreement until sometime between the rehearsal dinner and the first chord of the wedding processional). But the idea of conspicuous consumption has always been problematic for rich people, Veblen and Zahavi to the contrary. The problems arise on at least three fronts: mutability, mimicry, and signal inflation.

Mutability

The first problem is that there has never been any standard means of displaying wealth. Unlike house sparrows with their big black bibs or European barn swallows with their elongated tails, we have no universally recognized badge of status we can flash at suitable moments. The color purple used to be such a badge; it was the mark of royalty. (This may be why Richard Branson likes to wear a purple jumpsuit for his balloon expeditions, not the workaday orange of his support crew and copilots.) Purple dye once cost more than gold, because it had to be extracted from the vein of a shellfish, and it could take twelve thousand shellfish to get enough dye for a single dress or cloak. But purple is commonplace now. Gold may seem like an absolute standard, having served as the chief symbol of worth in Europe, the Nile Valley, southwest Asia, and India for more than five thousand years. But in China and in Central America until modern times, writes archaeologist Grahame Clark, jade was far more precious. When the Aztec leader Montezuma received the Spaniard Diaz del Castillo and graciously presented him with pieces of green stone to be carried home to his prince, Montezuma had to find a diplomatic way to explain to his disappointed guest that each piece was worth more than its weight in gold. In

Japan, which never regarded either jade or gold as particularly dazzling, people now put their diamonds on rings made of platinum, which is in truth the more precious metal. For that matter, diamonds only became a girl's best friend after the thirteenth century, when jewelers invented modern techniques for cutting them.

If there were a single defining feature for human symbols of wealth it would be this: They change. They change even when the symbols are physical features of our own bodies. Fat, for instance, customarily stood for wealth wherever famine and lesser forms of scarcity made food a status symbol. When the nineteenth-century explorer John Hanning Speke visited Karagwe in what is now northwestern Tanzania, for instance, he found that the king had his young wives force-fed on milk, swelling them to the point of immobility. Speke, a harbinger of Europe's burgeoning colonial sensitivity, actually had the milk-fattened wives rolled on the ground to be measured. One biologist has aptly likened these wives to honeypot ants, a caste of ants in an Arizona species whose nestmates pump them full of food until their abdomens swell to the size of cherries. Then they hang them up on the ceiling as living storage containers, "patient reservoirs of liquid sweetness," to be emptied again in hard times.

Corpulence also symbolized wealth in Europe and America until the early twentieth century. Being "a fat Adonis" like King George IV or "a corpulent voluptuary" like King Edward VII, served, in Veblen's terms, to demonstrate that the rich person was beyond the remotest need of ever doing useful work. For young, unmarried women, the slender waist continued to be the ideal. But, once married, the wives of the rich were frequently overfed and cooped up, like African queens and honeypot ants, for no purpose more challenging than the provision of domestic sweetness.

Even now, a few rich people keep up the fat cat tradition. The oil billionaire Marvin Davis is so hefty that he has an oversized chair specially designed for him at Spago Beverly Hills, where he often dines twice a day. When he goes to other restaurants, he sometimes brings his own chair with him (or to be precise, he has his bodyguards bring it). But fat stopped being a symbol of wealth at about the time it became possible for any wage-slave to achieve it. A 1960s study indicated that obesity was still largely an adornment of the upper classes in India, but a comparable study at about the same time in Manhattan found that only about 5 percent of upper-class women were overweight, versus 30 percent of lower-class women. Most rich people now aspire to be lean and muscular. They still practice conspicuous consumption of course, but instead of stuffing their bellies they do it by equipping their houses with exercise rooms like glorified machine shops. They pursue exercise as compulsively as they practiced gluttony a century ago: I have run across one centimillionaire and one billionaire who use the free time their wealth has secured them to spend no fewer than eight hours a day on some variant of the tread-mill—more or less what their ancestors did for minimum wage. In Aspen, the competition among multimillionaires for status often takes place on mountain bikes.

Fat and social status now go together only in one specialized form of display: An anthropologist at Texas A&M reports that "it is customary for upper-class parents in the Dallas–Fort Worth area to give their daughters breast implant surgery as high school graduation gifts. It is explicitly recognized by both parents and daughters that the young women will get more dates and be more popular in college if they have larger breasts. As one student put it: 'Among the wealthier families, the boys get hot cars for graduation, and the girls get big breasts.' "

Skin tone and hairstyle have also varied dramatically as badges of status. Most of us still like to show off a good tan in midwinter as if we've just gotten back from St. Bart's or St. Moritz, even if envious stay-at-homes mutter bitchily about skin cancer. In the debauched court around Louis XV, on the other hand, the height of fashion was to coat the face, neck, and bosom with powder to achieve an unwholesome paleness, accented by a *mouche* or beauty mark, as if one had just crawled out into the daylight after a weeklong *pas de deux* behind the drawn curtains of a canopy bed. Unfortunately, hairstyles in that era were so ornate, and so piled up on wires and scaffolds, that almost the only thing a fashionable woman could do in bed was sleep, sitting up. Fashionable men likewise often slept with a *bigoudi* on the upper lip to curl the mustache, another dismal triumph of display behavior over sex.

Mimicry and the Status Symbol Arms Race

The second problem with conspicuous consumption as a way of displaying wealth is that, like Christophe Rocancourt, almost anyone can do it, at least for a little while. At Jimmy'z Disco in Monaco a few years ago, the possibility of an appearance by Prince Albert attracted princess wannabes, and the wannabes in turn attracted men pretending to be princes. At one table, a group of five bachelors made nightly visits and rotated a single Rolls-Royce among them to improve their dating prospects. To a naturalist, this kind of disguise is reminiscent of the way dull, defenseless animals sometimes mimic species that are richer, prettier, or more deadly. Certain moths, for instance, go through life doing a brilliant imitation of a wasp (the stinging insect, that is, not the social class). Some beetles manage to pass for ants, and thus freeload on the ant colony's food, fancy accommodations, and security system. Humans do both; we are not so different as we like to think.

These disguises must be daunting for resource-conscious young women trying to sort out the sheik on the dance floor from the skank with the borrowed Rolls. In fact mimicry—the relentless co-opting of the signs and symbols of wealth—is the source of long-running warfare not just between men and women, but between the rich and almost everyone else. The upper classes "do not want anything from the Common Man," the current Duke of Bedford has written, tongue discreetly in cheek, "except that he should remain common. How can you shine if the common crowd is not common enough?" Middle-class mimicry obliges the rich to defend themselves "not only by staunch determination but also by machinations and trickery," and they are always dodging and weaving to keep the rest of us from catching up.

Thus it isn't an accident that the symbols of wealth change so often; it's a necessity. The code of conspicuous consumption must shift almost from day to day to keep the upwardly mobile from cheating their way in among the ranks of the rich. So the status symbol one year is Christmas in St. Bart's, another year it's the ranch in Patagonia. One year, a wealthy person's kitchen must have a $3,500 Sub-Zero refrigerator hidden behind wood paneling. The next year, Sub-Zeros become upper-middle-class, and the better kitchens have $10,000 Traulsens with clear glass doors (requiring staff to keep the leftovers inside looking pretty). This endless dance of move and countermove, the coevolution of imitators and the imitated, is at least as old as wealth itself.

A closer look at sumptuary laws, for instance, reveals that they were almost invariably attempts by the upper classes to limit conspicuous consumption by their imitators and keep the common crowd sufficiently common. Archaeologists, who have found widespread evidence for such rules, describe them as tools for "sanctifying, legitimizing, and separating the chief and other high status personages from the remainder of the population." A fourteenth-century petition to the English king, for instance, complained that "divers people of divers conditions use divers apparel not pertaining to their estate; that is to say, laborers . . . use the apparel of craftsmen, and craftsmen the apparel of valets, and valets the apparel of squires, and squires the apparel of knights . . . poor women . . . the apparel of ladies, poor clerks [use] fur like the king and other lords." The upstarts were actually driving up the prices indignant aristocrats had to pay for luxury goods. The clergy also joined the call to restrict extravagance, on the theory that it diverted money from the Church. And always underneath such practical considerations a sense of class outrage simmered at the unseemliness of such display by the unworthy. Agitating for new sumptuary laws in fifteenth-century England, one commentator fumed about how inappropriate it was for the lower classes to ape the wealthy fashion of wearing ankle-length fur-lined sleeves: "Now we have little need of brooms in the land to sweep away the filth from the street, because the side-sleeves of penniless grooms will gather it up." Yet the urge for upwardly mobile display was irresistible, as a seventeenth-century commentator acknowledged in complaining about the pretensions of the average merchant: "So long as he will see the tinsel of gold and silver glitter on the clothes of the nobles, he will rather mortgage all his worth, than not wear them."

Up against this powerful drive, sumptuary laws almost always failed, leaving the coevolutionary combat to be fought on the shifting terrain of the marketplace. Apart from frequently changing the symbols of wealth, one standard technique to fight off imitators has always been to make the displays bigger and more expensive: Build monster houses, buy better art, fly a fancier plane, and otherwise indulge in *really* conspicuous consumption. Thus a West Coast nouvelle riche whose family already has its own small air force recently remarked to a friend, "Did you know you can buy your own private aircraft carrier?" The idea was still in the talking stage. But with the Neiman Marcus catalogue having already offered a $20 million private submarine, capable of remaining submerged for twenty days, can personal Arleigh Burke-class destroyers be far behind?

Rich people often attribute this sort of excess to the pressures of the outside world. They feel compelled to make a big display, they complain, even though it goes against their own deeply modest personal preferences. A leading mergers-and-acquisitions billionaire, for example, had a vacation home on Red Mountain in Aspen that was merely comfortable. So he tore it down and had an 8,000-square-foot showplace erected on the same spot, ready to occupy by the following Christmas. "I used to have a nice little house in Aspen," he lamented afterward, rattling around in all that extra space, "and everybody else talked me into building *this*." He was undoubtedly being disingenuous. A man accustomed to getting his own way in all things does not demolish his house just because everybody else thinks it might be a nice idea. He does it to distinguish himself from the socioeconomic inferiors who relentlessly mimic his every move.

To discourage mimicry, the rich also continue to pass sumptuary laws in the form of environmental regulation. In Aspen, for instance, vacation homes of 15,000 square feet abound, and Prince Bandar, the Saudi ambassador to the United States, has a 55,000-square-foot weekend place. The real estate one-upmanship is so unrelenting that local authorities recently passed regulations limiting new houses to 5,000 square feet, though anyone who insists—and many people do—can still build a larger home by buying transferable development rights from property owners in areas where construction is outlawed. The town of Medina, outside Seattle, also recently imposed a temporary moratorium on houses larger than 13,500-square-feet, in part because monster houses put such a strain on the rest of the community. In the year 2000, when the average house in Medina used 80,000 gallons of water, for instance, Bill Gates's 65,000-square-foot compound used 4.7 million.

Every rich person is typically attended by a platoon of outsiders, as well as members of his own family, eager to alert him if he falls behind in the arms race with his imitators. The real estate agent, the architect, the interior designer, the couturier, the jeweler, the party planner, the art dealer, the yacht designer, the charity fund-raiser all conspire to crank up the pace of competitive displays. Behind his back, for instance, people sometimes mocked J. Paul Getty's ambitions as an art collector and gossiped that he liked a bargain better than a great painting. The art dealer Bernard Berenson did it to Getty's face. Determined to push him into a better class of art, Berenson once wrote Getty: "Thus far you have been getting a few pictures of the kind the Kress Foundation sends to every town where it has a five cent store." When a person becomes wealthy, gestures that might seem generous to the donor suddenly taste stingy in the mouths of the recipients. For instance, when Berenson died and an associate asked Getty to help endow Berenson's villa *I Tatti* in Florence as a center for art studies, Getty sent $1,000. The recipients took this as an insult and sent it back. Society expects the rich to make displays proportionate to their wealth.

Zahavi treats this idea as a biological principle: He argues that for any display to be persuasive, it needs to involve a significant cost for the individual making the display. That is, the cost needs to be significant not just in objective terms, *but from the individual's perspective, too.* Otherwise, no one takes it seriously. Most people understand this intuitively. A gift of costume jewelry can be deeply touching when it comes from one's grade-school child; less so from one's husband who happens to be a billionaire. Likewise, giving away $100 or $200 million a year made no material difference to Bill Gates in the mid-1990s, and so this philanthropic display didn't make much impression on the rest of the world, either.

Signal Inflation

This is the third problem with conspicuous consumption: Figuring out an appropriate cost and making a meaningful display gets increasingly complicated when you have enough money to gag a whale, or a pod of whales, even. How big a house can you build? How expensive a work of art can you hang in it? How rich a meal can you eat? How fabulous a diamond can you give your inamorata? Money becomes meaningless.

"Every million was less exciting than the one that came before it," one technology executive told me, though he was, to be sure, still out in the marketplace chasing millions. And a woman who has visited and lived in the most fabulous homes in the world confessed dispiritedly that nothing really impressed her anymore. "You become numb," she said.

Within the international community of the rich, it often seems as though everyone can afford almost any extravagance. Many perfectly good forms of conspicuous consumption thus lose their value. They suffer from the process Zahavi calls signal inflation: If everyone is rich enough to display a particular signal, or if "the cost of a signal is reduced to the extent that every individual can use it equally well, then the signal can no longer reveal differences in the quality or motivation of individuals." He cites the example of satin bowerbirds in Australia. Under natural circumstances, males of this species like to decorate their bowers with blue feathers, which are rare, and blue flowers, which need frequent replacing. Males compete by stealing blue objects from one another and by destroying their rivals' bowers. But near human habitation, blue plastic debris is appallingly abundant. For the bowerbirds, it must be like Croesus seeing his whole world turn to gold. They cannot quite break the habit of picking up blue objects, but they do so with none of their former zeal. Instead of stealing blue objects, they devote all their competitive energy to demolishing one another's bowers.

Among wealthy humans, lace is a classic instance of the most flamboyantly conspicuous consumption being destroyed by signal inflation. Until the mid–eighteenth century, lace-making required painstaking hand-craftsmanship. Its beauty and expense—the Veblen effect, again—made it one of the chief means for displaying wealth, and lace collars fanned out around the necks of the nobility like the ruff on a particularly magnificent bird. The width of the collar, strictly regulated by sumptuary laws, was at times a precise measure of social status. To this day, as a result, the Germans have an expression to put social inferiors in their place: *Du bist nicht meine kragenweite*, or "You are not the width of my collar." Hence also the abundance of sixteenth- and seventeenth-century portraits in which powerful men and women appear, as a French contemporary put it, in collars "gadrooned like organ-pipes, contorted or crinkled like cabbages, and as big as the sails of a windmill." This kind of display frequently tripped over from the sublime to the ridiculous, and not just in our eyes. A seventeenth-century caricaturist depicted one lady so enveloped in lace that she "looked like a turkey shaking its feathers and spreading its comb."

Even so, the effect was undoubtedly pleasing to the rich themselves who thereby declared their status unmistakably. It was pleasing enough to put up with starch, hot poking sticks for making pleats crisp again, and other inconveniences necessary to keep a collar in shape. Marguerite de Valois, wife of the French King Henry IV, wore such a large collar that she needed a spoon two feet in length for the perilous business of eating her soup. In England, Queen Elizabeth wore the highest and stiffest collars on the continent, often enriched with gold or silver lace. She introduced starch to her kingdom, another innovation of the rich, when she discovered that her coachman's wife had brought the secret with her from Flanders. She also posted guards at the London city gates to cut the ruffs of citizens who exceeded the dimensions permitted to their class. The sums spent on this sort of display were astonishing, even omitting the effect of inflation. The French King Henry III once appeared at a state affair wearing four thousand yards—more than two miles—of pure gold lace. A British aristocrat, Lord Berkeley, filed a lawsuit alleging that his tailor had shortchanged him by eighty ounces of silver in the lace for a single new suit. Under James I, the British royal family spent £614 on laced linen for the birth of their fourth daughter, Princess Sophia, who lived only three days. (This does not count the cost of having lace-trimmed coverlets carved in stone on her monument in Westminster Abbey.) The playwright Ben Jonson was hardly exaggerating when he joked that the art of being a gentleman then depended on your ability to turn "four or five hundred acres of your best land into two or three trunks of apparel."

Yet this whole highly elaborated form of display behavior collapsed into dust after the eighteenth century, when the development of machines to manufacture lace made it possible to equal the handmade product at a price even ordinary people could afford. Like satin bowerbirds inundated with blue plastic, the rich suddenly lost their zeal for the intricate lacework they had formerly coveted. "In such a case," writes Zahavi, "the signal loses its value. Because the signal's cost has gone down significantly, the signal is no longer useful and will disappear." The result is that even dinner tables seldom wear lace now, and the phrase "lace curtain Irish" persists as a slander on people too far down the socioeconomic ladder to realize that their favorite status symbol is now worthless. What happened to lace has of course also been repeated endlessly with almost every other form of display behavior throughout the industrial era. The pattern of original display being echoed by mass reproduction and almost immediately rendered worthless now happens nonstop.

The only thing that remains unchanged is the perennially ferocious human instinct for showing off. So what is the erstwhile conspicuous consumer to do? One possibility is to seek status by doing risky stuff. Larry Ellison of Oracle may no longer feel pure terror at losing billions in a single day, nor much joy at getting them back again. Such fluctuations can make no material difference to his quality of life. In his world, money itself has suffered severe signal inflation. But the one cost that always has meaning is a person's own neck. This may be a reason some rich people do such dumb, dangerous, and deeply impressive things—why Ellison wages mock dogfights in his Marchetti jet fighter; why pudgy Michael Price, the mutual fund superstar, plays breakneck polo well enough to get rated two goals on a scale of ten; and why Steve Fossett is perpetually falling out of the sky. Risking your life is always an effective display. This most conspicuous act of consumption still makes even jaded rich people pause and say, Oh.

And for the armchair rich? They often flock to objects associated with historical figures. Death insures against signal inflation because, as they say about waterfront property, they aren't making any more of it. This sort of badge of status also invokes the spirits of great men and women past, whom the rich may privately regard as their only true equals. One time, for instance, I visited Noelle Campbell-Sharpe, a self-made millionaire, the publisher of *Cosmopolitan*-style magazines, at her home outside Dublin. A maid escorted me to the sitting room, where there was a fire in the hearth and, on a table by the door, a copy of *Debrett's Distinguished People of Today*, with the page-marker left for easy reference at the entry for Campbell-Sharpe. A few minutes later she hustled in from work, a middle-aged child of the 60s with wild yellow hair and bright, slightly manic, eyes. After we'd talked for a while, she led me upstairs to a room she'd converted into a shrine to Napoleon. The ceiling was tented in green and gold silk, and the rug had Napoleonic bees in the border and a Napoleonic eagle at the center. She pulled down a book that was once part of Napoleon's private library and, breathing deeply, declared, "There are little pieces of Napoleon all over it." In rather the same vein, a plumbing manufacturer once advertised high-end fixtures with this inspirational copy: "Plato. Shakespeare. Mozart. They all went to the bathroom." You can, too.

The single most effective remedy for signal inflation is to confine one's display behavior to areas still susceptible to the most conspicuous expense—jewelry, houses, art, and philanthropy. The rich become connoisseurs in these areas, the better not just to collect for themselves but also to judge the merits of their rivals' collections. A woman may show up at a private party in Palm Beach, for instance, with what looks like a million-dollar diamond. "But she only wanted to spend $140,000," says a Worth Avenue jeweler. It is the equivalent of the long tail glued onto a barn swallow. "Maybe it has a little color. It isn't a fine white diamond. So I may caution the owner not to wear it during the day." Socialites frequently take courses in jewelry the better to sort out such dubious pretenders from the really "important jewelry"—"important jewelry" being a stock phrase among rich women, along with "substantial men." One day, for instance, I heard a rich woman remark on a wealthy Saudi family who adorned their five-year-old daughter with "pear-shaped diamonds, D flawless."

I expressed puzzlement. "D flawless," she explained, "are the finest white diamonds. D, E, and F are good diamonds. G and below, that means you're poor. With a trained eye, you can immediately tell. You cannot necessarily differentiate D and E without a loupe, but the prices are similar." Actually bringing out a loupe to study the diamonds on a kindergartner in the Saudi royal family would of course be bad form, the opposite of conspicuous consumption—conspicuous scrutiny. But taking courses in jewelry facilitates shrewd appraisals at a glance. Failing that, some people will actually stop to visit the jeweler on Worth Avenue: "They ask 'Was that really from your store? How much did it cost? How many carats?' " The store prefers to leave such revelations to the owner, who is often only too happy to oblige, and no doubt also exaggerate. "I don't sense a great deal of understatement," the jeweler said contentedly.

Philanthropic display gets the same finely graded scrutiny. It is not merely a matter of giving large sums but of giving them to the right organizations. Newcomers to Palm Beach often start by working with the opera, for instance, until they realize this does not get them invited to the better parties. The opera doesn't care who they are, says one local observer, "as long as the check clears." So people soon trade up to cancer, which is a B-list charity. "You see so much trading up in this town." With patience they may ultimately arrive at the A-list of the Crippled Children's Society (formally known as the Rehabilitation Center for Children and Adults) or the Preservation Foundation of Palm Beach (dedicated to the noble philanthropic cause of keeping Palm Beach charming for the rich). Likewise, the newly rich flock to art, realizing, as one Aspenite put it, that "if you don't have good art on the wall, that's a big status liability. You would not go there." Often, the newly rich opt for the most conspicuous possible art, or what a Paris dealer calls postcard art: " 'I want a Monet. I want a Monet that everybody's going to recognize. I want water lilies. I want haystacks. Or a Renoir. A girl at the piano.' So that everyone who walks in will immediately recognize that it's a work of great financial value." This sort of art is the most expensive form of conspicuous consumption, and it seems to promise the buyer instant ancestry, the appearance not merely of having arrived at the top but of having been there all the time.

What's Wrong with Conspicuous Consumption?

Eventually, though, it dawns on many rich people that even this is not getting them the social standing they seek. Something about the conspicuous consumption model does not work as advertised. Among animals, the most extravagant display is typically the best indicator of quality in an individual. Among humans, the need to be too conspicuous about wealth, intelligence, or almost any other desirable trait sends the opposite message, that there's something wrong, something a little off, about the person making the display. An overly lavish show of jewelry, like the one on my countess, for example, frequently stands in for lost youth and beauty. In her novel *Love in a Cold Climate*, Nancy Mitford described her fictional aristocrat Lady Montdore receiving guests in the splendid setting of her country estate: "Nobody could deny that on occasions of this sort she was impressive, almost literally covered with great big diamonds, tiara, necklace, earrings, a huge Palatine cross on her bosom, bracelets from wrist to elbow over her suede gloves, and brooches wherever there was possible room for them. Dressed up in these tremendous jewels, surrounded by the exterior signs of 'all this,' her whole demeanour irradiated by the superiority she so deeply felt in herself, she was, like a bullfighter in his own ring, an idol in its own ark, the reason for and the very centre of the spectacle."

But no one would want to marry her. "It did not seem to me that Lady Montdore could be described either as good-looking or as well-dressed," Mitford wrote, "she was old and that was that." To be fair, the purpose of this kind of display isn't to attract a mate; it is literally to dazzle, to induce awe and a certain giddy attentiveness, akin to flirtation, among the acolytes fluttering mothlike around the idol. "Important jewelry" can still make a woman of a certain age the center of attention, but it is also a diversion from the difficult truth that other, more important kinds of beauty have faded.

In truth, conspicuous consumption can be hazardous even for the young and beautiful. Consuelo Vanderbilt was a guest of the Hôtel de Paris in an earlier, less innocent, era. She quickly discovered that she was not supposed to recognize the men dining there in the company of beautiful women, even though some of the men had been her own suitors just a few months earlier. Vanderbilt, an eminently levelheaded and persistent character, extracted the reason from her pigheaded new husband, the Duke of Marlborough: The beautiful women were courtesans or, to use the debased modern term, hookers. Vanderbilt recounted the fierce rivalry between two of these women. La Belle Otero was "a dark and passionate young woman with a strong blend of Greek and gypsy blood, who was always flamboyantly dressed to set off her magnificent figure." Liane de Pougy was "lovely" in a more elegant way.

Their duel was fought with the lapidary weapons that were the chief prize among courtesans, then as now. "Fortunes were spent and lost for them and bets were exchanged on the relative value of their jewels," Vanderbilt wrote. "It was not surprising therefore that Otero should challenge her rival by appearing at the Casino one night covered from head to foot with priceless jewels. It was a dazzling display." The *demi-monde* twittered about how Pougy could possibly reply. Pougy, understanding that less can be more, showed up the next night in a simple white gown without a single jewel to distract attention from her D flawless figure. And, less being more only up to a point, she also brought along enough jewelry to turn Otero's lavish display to dross—and had her maid wear it. "In seeking to outdo her rival Otero had sacrificed good taste and had lent herself to ridicule," Vanderbilt concluded. The moral about conspicuous consumption was unmistakable. The clever and discreet Liane de Pougy ultimately became a *grande dame* by marriage to the Romanian Prince Ghika. Otero gambled away her fortune and died poor.

Inconspicuous Consumption

The art of being rich isn't, after all, about conspicuous consumption, but about *in*conspicuous consumption. Or, to give Veblen his due, it's about conspicuous consumption discreetly, even covertly, displayed. This is no longer strictly natural history, but a zone of perpetual combat between nature and culture. Our natural impulse is to show off with the splendid abandon of a peacock in his finest plumage: Look at my diamonds, look at my Degas, look at me. Vanderbilt, for instance, had the honesty to admit that she envied the courtesans. It seemed unfair to her that "the respectable woman, as she was called," had to refrain from showing off her beauty: "Any extravagance of fashion was condemned as bad taste, and no well-bred woman could afford to look seductive, at least not in public." Vanderbilt, a granddaughter of the commodore, was third-generation rich and, much as she might dislike it, understood perfectly well that an excess of "embellishments would immediately have committed one to the world of the déclassé." Yet she still struggled with her natural instinct for showing off.

This struggle is infinitely more difficult for the nouveaux riches, who have finally gotten their hands on money and want everyone to know it. "With an immense astonished zest they begin shopping," wrote H. G.

Wells in his 1909 novel *Tono-Bungay.* Of one character he adds, "So soon as he began to shop, he began to shop violently . . . he shopped like a mind seeking expression, he shopped to astonish and dismay; shopped *crescendo*, shopped *fortissimo, con molto espressione*." The nouveaux riches pile up jewels, cars, boats, town and country houses, and the attendant staff. "They plunge into it as one plunges into a career," Wells wrote. "As a class, they talk, think, and dream possessions." But gradually it dawns on them that, if the intent is to acquire not merely possessions but status, then they have probably chosen the wrong possessions.

"In 1981 I bought a Rolls-Royce. I wanted the world to know I was successful," entertainment lawyer Alan Grubman recently remarked. "Then in 1990 my dear friend David Geffen said, 'You know, Alan, it is not necessary for you to drive in a Rolls-Royce anymore. Will you please get rid of that car?' It was like I had a blinking neon sign on my forehead: NOUVEAU RICHE. The NOU went on, then the VEAU went off." Even Grubman himself now understands that the Rolls was a déclassé embellishment.

What's not so clear is why human culture should have evolved in this fashion, away from our natural urge to show off and toward less flamboyant plumage. In Consuelo Vanderbilt's case, mate-guarding was obviously one factor. Like a Muslim imposing the chador on his women, the upper-class husband of Vanderbilt's day wanted to limit the public seductiveness of his wife as a way to minimize her chances of being unfaithful. That way, he could be more certain of the paternity of their offspring, and thus have peace of mind when he headed off to lavish baubles on his courtesan in Monte Carlo.

Security in the larger sense also argues for inconspicuous consumption: The way pretty women attract playboys, people who display their wealth too publicly risk becoming targets for thieves, kidnappers, social satirists, fund-raisers, and the scolding, solicitous clergy. At the very least, resentment from her neighbors, or, perhaps even worse, sheer incomprehension, is the common fate of a woman who wears a Jean Paul Gaultier original in Clarksburg or Prairie du Chien. So the rich often learn to practice a dual survival strategy: Inconspicuous consumption in the hometown that actually made them rich, combined with a club, social set, or vacation getaway in one of the enclaves of the pseudo-species, where it is socially acceptable to step up the level of display.

A similar strategy of concealment and display also occurs in the natural world: There are, for instance, lavishly ornamented butterflies and moths that disguise themselves as bits of tree bark or as dead leaves or, in certain truly shabby-genteel species, as bird droppings, the better to avoid detection by predators. Some moth species only flash the brilliant colors of their underwings to startle predators, and some butterflies only show their true colors when they are fluttering around among their own kind. They roll out their wasteful sexual displays mainly when there's a chance of attracting a desirable mate. They send signals recognizable only to each other. Female moths, for instance, don't broadcast their amorous intentions to the sweating mob. Instead, they send out a delicate perfume, which can be detected only by suitable males of the same species. The male's elaborate antennae, each quivering with 1,700 sensory hairs, are utterly attuned to detecting this telltale scent. The female in turn may judge by subtle nuances in the male's aroma whether he is a worthy mate.

Being rich must sometimes feel like that. That is to say, the rich also frequently decide that it is safer, or simply more pleasant, to travel incognito. When he was a student at an American college in the 1970s, for instance, Prince Albert of Monaco tried to fit in by calling himself Al Grimaldi. For roughly the same reason, Kenneth Thomson is largely unknown in his hometown of Toronto, though he is among the richest human beings on Earth, and also one of the world's most powerful newspaper barons. He does not want to have his picture taken, he once told a photographer, because he likes to be able to pop over and buy socks on sale at the department store across the street from his office tower. This when he was himself the owner of the department store. At the same time, the rich generally want to be able to flash their underwings at people they hope to impress, and they tend to have all their vibrissae (or at least the vibrissae of their personal assistants) aquiver to sort out which people are really worth impressing. Hence the elaborate system of references to schools, restaurants, resorts, and famous friends, which serves as a code, like the moth's subtle perfume, for sorting out the worthy from the un-. They dress down, but almost always with hints: The gabardine coat has a sheared mink lining, and the plain white blouse has the Burberry plaid on the *inside* of the collar, where only those the wearer invites close enough can see it. The work shirt looks like ordinary denim, but those who can afford this sort of thing might recognize it, by the lilac topstitching, as a $340 product of Borelli, the world's best shirtmaker.

Hence also places like Monaco, Palm Beach, and Aspen, which exist, in truth, largely as safe havens for rich people to come together and indulge the powerful urge to display that is so firmly repressed back home. The Worth Avenue jeweler noted that visitors from Boston "object the most" to showy jewelry, "but they buy the biggest." Palm Beach, another local confided, "is one of the last places in the world you can still wear important jewelry in public," and just to be sure it stays that way, police routinely check the license plates of cars coming across the bridges into town. Monaco likewise goes to extraordinary lengths to ensure that a woman can wear her best diamonds and safely wander home with them at 4:00 A.M., or even leave them on the front seat of the Bentley, with the keys in the ignition and the engine on, and find everything still there in the morning. The Monaco police have eighty-one video cameras trained on every street and public space of a nation the size of a small town. "A friend of mine was walking home one night, when she heard footsteps behind her," a woman told me, over champagne. "She started to walk faster, and the footsteps got faster, too. So she ran for an emergency phone, and a policeman answered, 'Hello, Madame Dubost. He's one of ours and he's escorting you home.' "

Security considerations aren't enough, though, to explain the mandate for inconspicuous display. Even within the safety of their own world, the rich consider ostentation déclassé. The people committing the ostentatious acts frequently attribute this to envy, if they notice it at all. That's the subtext of the often-told anecdote about Princess Margaret's reaction to the Cartier diamond worn by Elizabeth Taylor: "It's so big. How very vulgar." In reply, Taylor supposedly slipped the ring onto Margaret's finger and said, "It doesn't look so vulgar now, does it?" But while there are many reasons a member of the British royal family might envy Liz Taylor, big diamonds probably aren't one of them. By and large, old money sorts don't live in fear of seeing something that might make them envious.

The Artful Showoff

So is inconspicuous consumption just old money's gracious way of being modest about wealth? On the contrary. Being inconspicuous about wealth, being casual, or even indifferent to its splendors, is ultimately the most effective way of showing them off. When Katharine Hepburn used her Oscar as a doorstop in the bathroom, she wasn't being modest; modesty would have stowed the thing away in the attic, not in the one room every guest was likely to visit for a moment of quiet contemplation. But having won four Academy Awards (and been nominated for twelve), Hepburn could easily be casual about them. The casual style of the display, in such cases, often impresses at least as powerfully as the object on display. A certain offhand or blasé manner, like the one that came quite genuinely to my friend of *grande bourgeoisie* family in Paris, says that she grew up with Monets and Sisleys on the walls, the way other people grew up with scenes from the life of the Blessed Virgin as depicted in the annual parish calendar from Our Lady of Perpetual Sorrows. On visiting an English country house for a shooting weekend, my Paris friend was duly impressed to see a Holbein hanging in the hallway and, in the library, the biggest Stubbs she had ever seen. When she saw a huge painting of a young girl in the living room, she innocently exclaimed, "Oh what a pretty picture," and was again impressed when her host said, "Yes, that's grandmum painted by Sargent." What really demolished her sense of having seen it all before was when she went up to get ready for dinner and found a Gainsborough hanging in the dressing area of her guest bedroom. The effectiveness of the display stemmed at least in part from its seeming so natural. At dinner, a young woman she had seen earlier dressed in knee socks turned up' "wearing a gown with emeralds like this," and here my friend's pursed hand leapt rabbitlike along her collar bone, "*Choong! Choong! Choong!*" This turned out to be her hostess, who at the end of the evening wore the same gown, and a pair of wellingtons, to go out and tend a sick horse.

   "Decorating out of the attic," the interior design of established wealth, sends a similar message of being casual and at ease. The home-owner gets to sound a charming note of self-effacement as when Tara

Rockefeller, wife of Michael, recently said, "My mother calls our decorating style 'Early Attic, Late Cellar.' " It also announces that the happy couple has access to the attics and cellars of the Rockefeller family. This sort of understated display functions as one of the most effective barriers against the nouveaux riches, according to Nelson Aldrich, Jr., author of *Old Money* (and himself a poor cousin to the Rockefellers), because "you cannot buy into it." Old money distinguishes itself in such cases by consuming nothing. One of the Earls of Lonsdale apparently dressed himself, as well as his house, out of the attic. When a friend admonished him for his shabby appearance, he replied with perfect old money self-confidence: "In London, nobody knows who I am, so it doesn't matter. In Cumberland, everybody knows who I am, so it doesn't matter." Decorating out of the attic still functioned as a badge of status in England as recently as the 1980s, when Michael Heseltine, a self-made publishing millionaire, was a senior minister under Prime Minister Margaret Thatcher. "The trouble with Michael," an upper-class member of his own party sneered, "is that he had to buy all his furniture."

What's a little puzzling, if one is striving to be graceful and at ease, is why old money sorts often bristle with such anger and disdain toward conspicuous consumers. Conspicuous consumption of course marks an individual as a newcomer, and in the animal world, an individual entering a new group risks suffering suspicion, abuse, and even death. In one study in Rwanda, females entering a large group of mountain gorillas experienced harassment from resident females in the form of lunges, chases, shoving, screaming, and the occasional "pig grunt." With the studied crassness of the *nouveau riche*, the harassed newcomer sometimes responded with nothing more than a "belch vocalization." The harassment was frequently "spiteful"; that is, the individual doing the harassing didn't get more food or other benefits for her trouble. She seems merely to have been delineating the boundaries between in-group and out-group. In other species, including feral horses, yellow baboons, and ring-tailed and brown lemurs, female harassment can delay pregnancy among newcomers. Outsiders represent a threat to individual status and to the stability of the group. Harassment minimizes or delays the threat by making it harder for newcomers to fit in.

The threat alone doesn't by itself satisfactorily explain why ostentatious display is such a taboo. The people at the top of the hierarchy are frequently far too secure to feel any threat or too shrewd to show it when they do. Moreover, even friends and social allies find public displays of conspicuous consumption vexing. Why should David Geffen have been so irritated—"Will you please get rid of that car?"—by his dear friend Alan Grubman's Rolls-Royce? After all, Geffen was himself nouveau riche, at least chronologically. (By Hollywood standards, having a fortune that dates back twenty or thirty years is like having had ancestors at the Battle of Hastings. But we are still talking about wealth of post-Nixonian vintage.) The problem with the Rolls-Royce wasn't merely that it represented such a naked attempt to show off his wealth, but that Grubman, by his own admission, "wanted the world to know." And wanting the world to know is something a rich person ought not to do; it is an affront to other rich people. According to the peculiar code of the pseudo-species, proper display behavior for a rich person should be aimed principally, if not exclusively, at other rich people. Discussing the custom of decorating out of the attic, for instance, Nelson Aldrich puts it nicely: "The whole point of inculcating the peculiar aesthetic of the class is to lift its habitat above the quick and nasty transactions of the cash nexus to the exalted plane of disinterested delight."

It's not that other people don't count. They count negatively. A display may actually lose status to the degree that it is comprehensible by the untutored masses. On this and other grounds, a Gerhard Richter outranks a Norman Rockwell. A game of bridge outclasses five-card stud. The best displays speak in a private language known only to other rich people. A fashionable woman might wear a brooch that looks to most eyes as if it's made of steel. But the right people will know that, in fact, the owner traveled to Paris to buy it, to a small shop on the Place Vendôme which appears to have gone out of business, with faded mauve velvet in the windows, and no jewelry on display, nor any sign except "J.A.R.'s" etched in mirrored glass overhead. There, over a desk covered in worn blue leather, a jeweler named Joel A. Rosenthal meets privately with the richest women in the world and practices the peculiar art of making platinum look like steel, and pink sapphires pass for everyday amethyst. His work is considered art, but only other people who can afford it would guess that a typical piece can sell for $30,000. Limiting the audience in this way is the point: "To the individual of high breeding it is only the more honorific esteem accorded by the cultivated sense of the members of his own high class that is of material consequence," Veblen wrote. He argued that the wealthy leisure class had grown large enough "to constitute a human environment sufficient for the honorific purpose." An environment, in other words, suitable for the pseudo-species of the rich, in which "there arises a tendency to exclude the baser elements of the population from the scheme *even as spectators whose applause or mortification should be sought.*" For the rich, that is, wanting to impress a stranger on the street makes about as much sense as a peacock wanting to impress a dog.

The display behavior of many rich people thus evolves toward increasingly inconspicuous consumption. Not the obvious painters, but more obscure and intriguing ones; not the postcard pieces, but others that somehow distill the essence of an artist. The rich become collectors of prints that are too fragile to hang on the wall, books of hours that are not susceptible to being shown off at a cocktail party, and brooches with the diamonds hidden in back. This may not seem like display behavior at all. A piece may come out of its hiding place only once every few years. But this merely makes the display more selective, more potent, by implying both that the owner has limitless resources and that the lucky viewer is among the one or two other people in the world who might truly appreciate it.

The Holy Grail

In the course of my research, I made a point of asking every person I interviewed to name the single most impressive display each had run across—something so costly, so unique, so exquisitely tasteful, so intrinsic to the identity of the person displaying it that no one could possibly hope to imitate it. This question almost always stumped them. People who seemed to me to devote their every breathing moment to the veneration of material embellishments seemed, to themselves, never to think about display at all. Perhaps my question smacked of the cash nexus or, worse, of the monkey's red arse. And they were aloft on the exalted plane of disinterested delight.

So here is my own *choong-choong-choong* moment. One afternoon, I visited Prince Rainier at his office in the Grimaldi palace in Monaco. Given that he is himself an aristocrat, known formally as S.A.S. Prince Rainier, for *son altesse sérénissime*, his most serene highness, I was a little surprised to see a scene from the French Revolution hanging in his foyer. Rainier, whose power depends on the indulgence of neighboring France, gave me a quick grin and said, "It's good for my French visitors to see." A caveat on entering.

Later, as I was leaving, I noticed a painting hung unobtrusively over the door. I recognized the hazy image of the harbor, with the Grimaldi palace atop the Rock overlooking the Mediterranean. Prince Rainier mentioned that the moored yacht in the painting had belonged to his grandfather, Albert I. The painting was thus an anthology of uncheatable status symbols: a palace where Rainier's family has ruled for 700 years, a nation the family effectively owns, and grandfather's handsome yacht. For the visitor who has taken the caveat on entering, the message, on going away, is about family permanence as unshakable as the Rock itself. It is remotely possible that some nouveau riche, someone whose family power dates back, say, a mere 300 years, might somehow buy this painting. But no one could ever acquire the impeccable provenance.

The style of the painting looked familiar, and I asked Prince Rainier the identity of the artist.

"Claude Monet," he said.

9

Living Large

The Habitats of the Rich

If they could see me now,
My little dusty group,
Traipsing 'round this
Million-dollar chicken coop!

—Dorothy Fields, *"If My Friends Could See Me Now"*

ONE AFTERNOON IN THE MAROON CREEK NEIGHBORHOOD ON THE outskirts of Aspen, a developer was showing me around a house he had recently sold. The new owners, who paid more than $10 million for the place, had settled in, and it looked like home. A very rich home at that, with 9,000 square feet of space encased in fifteen-inch logs, a chandelier made of nested elk antlers over the front entry, a collection of silver-topped canes in an ornately carved umbrella stand in the front hall, and beyond, across the vast living room, a floor-to-ceiling view of ski slopes and forested mountainside.

"We work our tails off so when you open the door you've got a million dollar view," said the developer. "You say, 'Hey, I'm in the mountains.'" He was a bullet-headed, blue-eyed man, in black turtleneck and leather jacket, with a rat-tat-tat, time-is-money manner. This was a turnkey house, which is to say, a time-is-money kind of house. "The big reason people buy turnkey is time," he said. "Time is the most valuable commodity. Most of these people don't want to take on a *project*, especially not in a resort community away from their main base." So he makes it his business to sell them the finished product with all the signals of wealth-in-Aspen already built in.

Thus, despite the homey appearance, hardly a single item on display, not even the pillows embroidered with cute sayings, was intrinsic to the identity of the rich and semifamous people living here. The buyer had merely paid the price and turned the key in the door. The construction, the furnishing, the accessorizing, the taste, were all the developer's, down to the decorative chests, the little silver-framed pictures, the candelabras, the towels (but not the robes) in the bathroom, the Calphalon and Williams-Sonoma in the kitchen, the coffee table books, and the leather-bound volumes on the mantle, including one titled *Lisez-Moi*, which is apparently how you say "fat chance" in Paris. The developer had even chosen the paintings on the walls. "Is it collectible? Is it Schnabel? No," he said, bolting past the handsome frames and generic Western scenes. "The paintings run from $2,500 to $8,500. They're more than wallpaper." A huge trophy head presided from high over the fireplace. "I put a moose in about half the houses I do," he said. "Nice animals. I don't do anything controversial. I just try to do something that says 'Mountains.'" Preferably in a deep, masculine bellow.

The idea of the turnkey house, generic housing for the rich, is odd, and it is a stretch to suggest that there is any precedent in the natural world for this kind of thing. Even so: When yellow warblers arrive in Jackson Hole, Wyoming, each spring, the trees are still leafless and there's snow on the ground. But the warblers seem to recognize, by the shape and color of the bare willow brush, that this is good habitat for yellow warblers. Likewise, when scientists offer cage-reared chipping sparrows both deciduous and coniferous branches, they perch on the coniferous ones, which they are genetically prepared to recognize as a cue to their ancestral habitat.

The turnkey developer is merely in the business of supplying habitat cues to the rich—knotty alder cabinets, heavily distressed, in lieu of bare willow, three-inch-thick Italian granite countertops in lieu of pine tree

Marilyn Monroe and a chimp deploy variants on lip-smacking, a primate appeasement gesture. Unlike a ruffed lemur, the rich can change their badges of status at will, though the effect is sometimes alarmingly similar.

For J. P. Morgan, John D. Rockefeller, S
and a dominant mandrill, projecti
ferocity with a determined glower was
way to make rivals step aside.

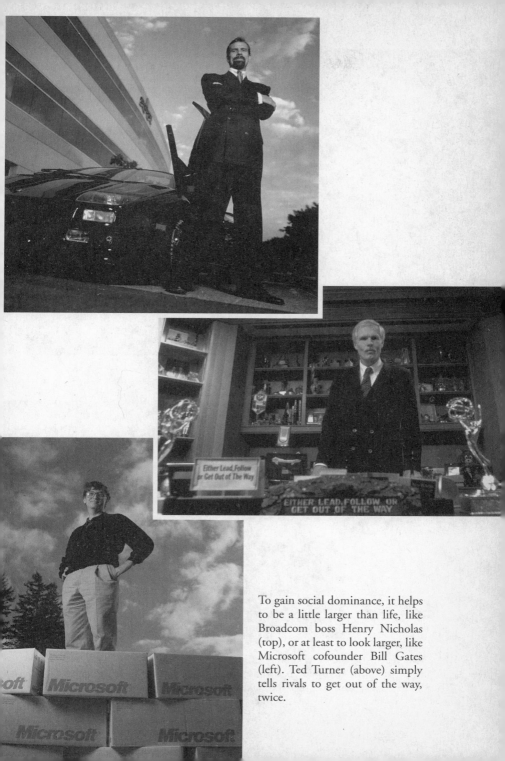

To gain social dominance, it helps to be a little larger than life, like Broadcom boss Henry Nicholas (top), or at least to look larger, like Microsoft cofounder Bill Gates (left). Ted Turner (above) simply tells rivals to get out of the way, twice.

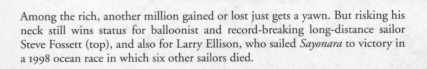

Among the rich, another million gained or lost just gets a yawn. But risking his neck still wins status for balloonist and record-breaking long-distance sailor Steve Fossett (top), and also for Larry Ellison, who sailed *Sayonara* to victory in a 1998 ocean race in which six other sailors died.

An undercurrent of ferocity runs through the lives of the rich. Blenheim Palace is a vast trophy case celebrating conquests both military and sexual. A penchant among the rich for leopard or snakeskin makes the wearer seem dangerous, and evokes an innate predator response in other people. But few rich people have taken fierce fashion as far as the Indian prince who celebrated his might with a sculpture which both growled and screamed, one way of reminding his guests to behave.

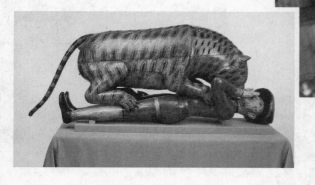

The polygyny threshold suggests that dominant males—whether bull elephant seals or the likes of (clockwise) Donald Trump, Gordon Getty, and the Marquess of Bath—tend to attract more females. Edward VII, one of the great polygynists of the Victorian era, advertised his prowess with a manly pose.

Wealth can give greater sexual freedom to women like (top left) Pamela Harriman and Jennie Jerome, on whom the bluebird of happiness seems to have settled with her special knack for infidelity. Asked how many husbands she had had, the heiress Peggy Guggenheim (above) once replied, "My own, or other peoples'?"

The offspring of the rich never know where the advantages of family leave off, and their own accomplishments begin. Hong Kong businessman Richard Li gets plenty of kin selection from billionaire Dad Li Ka-Shing, but refers to "father" as "the f word." Like Alva Aldrich and John D. Rockefeller, Jr., the offspring of the rich tend to marry one another, a process biologists call "assortative mating." But they can end up, like Consuelo Vanderbilt, as just "a link in the chain."

branches—and he could almost be talking about yellow warblers instead of multimillionaires when he says, "Time is such a critical asset to these people. They just want it to be done. So it's an emotional buy." Nest here, nest now. "Plus, it's a sound fallback investment strategy. Can we get our money back out? You bet. And then some."

Like other animals, the rich select their habitat according to specific cues, mainly cultural ones like the Steinway concert grand in the corner or the knowledge that a Ford or a Bass lives down the street, but also biological ones, like the presence of water, large trees, open space, and distant views. These biological habitat cues date back to the savanna environment in which early humans evolved: We love waterfront property because it comforts us the way proximity to a watering hole once did. Failing waterfront property, we like clean, bright surfaces because they evoke the glitter of water. (Dragonflies sometimes get fooled in the same way, laying their eggs on the hood of a car instead of the surface of a pond.) We take comfort in a certain branchy tree shape because our simian forebears roosted in such trees at night, for safety from predators. We all apparently respond to these same habitat cues. But rich people can actually afford to own them.

Blenheim

When the rich settle down to build their dream houses, what they often seem to have in mind is something like Blenheim Palace, only better heated. Blenheim is an extraordinary mix of both cultural and biological habitat cues, and it's also a useful study in the tangled relationship between a house and the family it shelters. I saw it for the first time just before sunset on a spring evening. The main street in Woodstock, a small town a few miles north of Oxford, England, swings left at the end into a cul-de-sac enclosed by a high stone wall. A narrow gate in the middle of the wall was still open and, walking through it, I felt as if I had crossed into another dimension: Off to the right, in the middle of a vast private park, a lake perfectly mirrors the triple arches of a glorious stone bridge. (Or as Lytton Strachey put it, on this same spot, "There is a bridge over a lake which positively gives one an erection," a response which must be taken, given the speaker and the setting, for submissive display, not dominance.) Beyond, in the distance, stands the palace itself, a mass of columns, arches, and towers, the stone glowing like honey in the evening sun, golden spheres on the rooftops glinting against a bruised April sky.

It is arguably "the finest view in England," as Randolph Churchill, whose family still owns the place, once boasted on this spot, to his new bride Jennie: Three acres of house (seven if you count the courtyards), set in a 2,100-acre park landscaped by Capability Brown, enclosed by a stone wall nine miles long, altogether so grand that King George III once remarked, "We have nothing to equal this."

Grandeur and envy being near of kin, Blenheim has also been one of the most reviled buildings in all of England and by some of the most scathing critics, among them Alexander Pope ("the most inhospitable thing imaginable, and the most selfish . . ."), Horace Walpole ("execrable within, without, and almost all around"), Voltaire ("*une grosse masse de pierre, sans agrément et sans goût*"), and Noel Coward ("Woke frozen. Shaving sheer agony . . . loo like an icebox . . . [duke] none too bright").

The Churchills themselves, or at least their wives, have at times denounced Blenheim Palace as "the dump" and "that wild, unmercifull house," and with good reason. No house anywhere has ever put its mark so thoroughly on a family's collective soul. Built between 1705 and 1720 to glorify John Churchill, the first duke of Marlborough and one of England's greatest military heroes, Blenheim instead helped ruin his reputation with its sheer egotistic excess. Built to found a dynasty, Blenheim instead turned succeeding generations into madmen, scoundrels, snobs, and gold-diggers, the better to keep the roof overhead and to gild the plasterwork lilies. Yet the arrogant mythology of the place also worked. If the owners were, at best, ordinary men and women, it was often hard to notice, given the ostentatious way every stone declared their dominance. And at least once in 300 years, Blenheim also redeemed itself by producing someone worthy of the place. Winston Churchill, one of the greatest statesmen of the twentieth century, was born at Blenheim, the grandson of a duke, and he got his sense of destiny there. "We shape our dwellings," he once wrote, "and afterwards our dwellings shape us."

Full Larders

Ecology, the science of an organism's relationship to its surroundings, derives from the Greek word *oikos*, or "house." The first thing to be said about the ecology of the rich is that their houses are so very big. Blenheim, for instance, has 187 rooms, plus another 33 in the bridge across the lake. The houses of the rich have been big almost as long as there have been rich people to live in them, for what was in the beginning an eminently natural reason: food storage.

Human wealth didn't merely originate in agriculture. It also consisted until recent times of surplus food. This idea still threads through our language about wealth. Money is of course "bread" or "dough," and though the word "pecuniary" suggests lofty dealings by international finance types sequestered at Davos, it was originally about counting heads of cattle, or *pecus* in Latin. A socialite on a "spree" at Bergdorf Goodman is likewise indulging in what the Scots called a *spreath*, or cattle raid, though she may now be hellbent only on Gucci leather. There is, alas, no cow connection in "moolah." But "plutocrat" derives from the Greek god Pluton, who was thought to be pushing up grain, the wealth of the ancient world, from underground.

Private food storage sneaks into the archaeological record of the Near East beginning about 9,500 years ago. Before then, early agricultural settlements apparently practiced communal storage of any surplus food. The first private storage consisted of little more than a stash in the corner of a room, like a squirrel's hoarded acorns, or the choice morsels hidden from his troop by a chimp. But architecture itself was embracing big new ideas then, including foundations, walls, and doors, not to mention rooms with corners: Rectangular houses replaced circular huts as agriculture encouraged people to settle down closer together. The houses gradually got bigger, with storage cubicles around the periphery, and then they headed skyward, with the second floor apparently devoted to living space, and the first to storage of food. Houses developed courtyards, doubtless as places to keep the livestock safe from plutocratic neighbors off on a spree. At some sites, such as Çayönü, a Turkish village which flourished 10,000 years ago, houses in the immediate periphery of cultic areas were bigger, better built, equipped with status objects—and clearly set off from the rest of the community, the beginnings of the social isolation of the rich.

Bigger houses meant a bigger agricultural surplus, and thus more wealth. Of a big house at Loma Torremote, a 2,500-year-old site in Mexico, for instance, an archaeologist writes "the architecture itself indicates the family's economic importance. More than 2.2 times the family's annual grain needs could be stored in the pits within this compound, while the storage capacity of neighboring compounds indicates their annual needs would be just barely met."

Control of surplus food was still the basis for wealth in the prime of the English country house, on which many rich people model their homes to this day. In Ireland and England until the end of the nineteenth century, "the big house" dominated every rural district. The house typically controlled at least a thousand acres of land, and in many cases much more, representing sustenance for a fiefdom of servants, laborers, and tenant farmers. This territory served as the basis for all political power, since the lord of the big house typically controlled the district's representation in Parliament and also appointed lesser officials, down to the local vicar. As recently as the 1870s, when the British Isles had a population of 26 million people, just 7,500 families secure in their big houses still owned 80 percent of the land. Land holding was so important to rank that the nineteenth-century prime minister Benjamin Disraeli declined a dukedom on the grounds that he lacked sufficient land to support the dignity. As a vestige of this territorial spirit, the current duke of Marlborough, who owns 11,000 acres in Oxfordshire, still thinks of himself primarily as a farmer, though tourism now generates most of the income at Blenheim Palace.

The houses of the rich aren't, of course, about storing food anymore. Nobody devotes a wing of the house to an orangerie when citrus is for sale at any grocery; that's signal inflation. But they are still big. Gargantuan, preferably. They also still have plenty of storage space, though mainly for clothes now. Candy Spelling, wife of Hollywood television mogul Aaron, has a 3,500-square-foot closet devoted, she says modestly, to "real casual, run-around things." But lots of them. And when a New Yorker was building a 43,000-square-foot third or possibly fourth home, on the ocean in

Palm Beach, her main complaint was that the bedroom closet was so small. "It's as big as my living room," the architect protested, to which the owner replied, "Why do you live in such a small house?"

The sheer scale of these homes can intimidate visitors, which has of course always been the other great purpose in building big. Or as the Palm Beach woman put it, with semi-ironic frankness, as we sat one evening on her patio: "It says I've got it, and you don't, and you have to kowtow to me." (And, yes, OK, I did—but not so much as the manicurist who sat on the floor in front of her, round and sweet as a Botero figure, cradling her heel and telling her, "No, your feet are beautiful.") This home team advantage, times ten, or twenty, or a thousand, can be enough to make hapless visitors forget who they are. A woman who grew up in such a house recalled the experience of bringing her childhood friends home to visit: "It was interesting—the comments and the looks. Usually it was just silence. They would drive up the driveway and they'd get quieter and quieter . . . it was difficult for them to remember their names."

Female Coalitions

The dynasty at Blenheim got its start in 1666, when an impoverished Dorset gentleman named Winston Churchill, a staunch monarchist during the Civil War, secured a position in the royal court for his pale, gangling teenage daughter Arabella. There could of course be no dynasties of any sort without women. But maybe because Blenheim's history dwells so lovingly on two great men, John and the later Winston, it's striking how much the fortunes of the Churchills have actually depended on women. What's even more striking, if one happens to be reading the history of Blenheim and, at more or less the same time, a book like Frans de Waal's *Chimpanzee Politics: Power and Sex among Apes*, is how often male power in primates is determined by female coalitions. Watching his chimps at the Arnhem zoo, de Waal observed that alpha males were big on noisy bluster but held their ground mainly when they had the active support of a female coalition. It was the same at Blenheim.

Soon after arriving at the royal court, Arabella Churchill became, as one historian has put it, "the Duke of York's doxy." The duke was the younger brother of King Charles II and later became king himself, as James II. Arabella bore him several bastard children and apparently used her sway to get her younger brother John a place in the duke's retinue. John also quickly made his mark in the splendidly debauched circumstances of that court. He caught the wandering eye of the king's own mistress, Barbara Villiers.

According to the story historians generally credit, Villiers was entertaining Churchill one night in her bedroom at the site that is now 10 Downing Street, when King Charles came to call. Churchill leapt out the window, for which Villiers later thanked him with a lavish gift of £5,000. People still debate whether, as one local put it, "he got the money for jumping into her bed or out of it." Villiers herself never said, except to complain that she gave it "for very little service." But kin selection, rather than any form of sexual gratitude, seems a more likely explanation for this gift: John Churchill was Villiers's cousin and also the likely father of her sixth child. In any case, Churchill prudently invested the gift in a life annuity paying him £500 a year, and thus began the fortune on which the dynasty at Blenheim would be founded. Villiers also paid the standard bribes required to get Churchill valuable positions in the court and the army. A contemporary remarked that "a man who was the favourite of the King's mistress and brother to the Duke's favourite . . . could not fail to make his fortune."

But it was another member of the court, Sarah Jennings, who really set the Churchills onto the paths of glory. In her portrait off the main hallway at Blenheim, she appears as a beautiful young woman with fine, golden hair, rosy cheeks, a lower lip like a velvet pillow, and hard blue eyes. "Her with the fury heart and fairy face," as one poet put it. Sarah was a member of the royal court, but played her hand differently from Arabella and Barbara. She yielded nothing to John Churchill's passionate entreaties until he broke down and asked her to marry, an event their descendant Winston later described as "his only surrender." The marriage was a love match. But if surrendering to a woman with no family fortune was risky for such an ambitious young man, the choice also proved shrewd.

John and Sarah were, in fact, a thoroughly modern couple, always on the make. Sarah kept her job as an attendant to the Duke of York's daughter, Princess Anne, and was proud to earn her own money. On the surface, the job consisted of little more than tending to royal whim, trading gossip, and gambling at cards, Sarah often thriving at Anne's expense. As in a troop of bonobo monkeys, being close and performing little acts of grooming provided the understructure for important friendships. Sarah was beautiful and witty, and the hapless young Anne adored her. Anne's passion for Sarah was so intense, according to the historian Virginia Cowles, that it "must have had its roots in physical attraction, though the Princess herself was probably unaware of it." A contemporary remarked that "there never was a more absolute favourite in a court." What Sarah wanted, Anne eagerly gave her. As Anne's father the Duke of York became King James II, soon to be succeeded by King William III, and finally by Queen Anne herself, what Sarah asked for and consistently got was advancement for her husband, John Churchill. The couple was ultimately earning £65,000 a year from their various titles and positions—roughly $11 million in today's money.

Unlike many beneficiaries of royal largesse, Churchill actually earned it. As commander of the British forces during the War of the Spanish Succession (1702 to 1711), he defeated the army of King Louis XIV not once but ten times and never lost a battle. He was a brilliant general, adept at negotiating treacherous politics on the home front, attentive to the care and feeding of 100,000 soldiers in the field, and, above all, shrewd and surprising in battle.

Churchill's style of command kept him in the thick of battle, occasionally even at the head of a cavalry charge. Once, when he was mounting his horse, a cannonball passed between his legs and killed the man standing beside him. In another battle, one of his captains wrote that on the brink of a dubious attack, "The Duke of Marlborough (ever watchful, ever right) rode up quite unattended and alone, and posted himself a little on the right . . . whence he had a fair view of the greater part of the enemy's works. It is quite impossible for me to express the joy which the sight of this man gave me at this very critical moment." He was, on the other hand, a nightmare for France. For generations afterward, French schoolchildren believed that if they misbehaved, a monster named "Malbrouk," the Duke of Marlborough, would get them.

Churchill's crowning achievement was his victory in 1704 at Blindheim, Germany, which British soldiers pronounced "Blen'm." It was the first time the British had defeated the French since Agincourt almost three hundred years before. Queen Anne, no doubt encouraged by Sarah, chose to honor her triumphant general by giving him her royal park at Woodstock. Parliament enthusiastically agreed to build him a great house there, and Sarah made it her ambition to immortalize the Churchill lineage, her lineage, in the vast mausoleum to be called Blenheim.

Corpses in the Undercroft

Beyond the practical realities of power, the big houses of the British countryside customarily embodied a myth: Each became synonymous with the ruling family, whose ancestors were preserved in oil paintings on the walls of the house, and often buried in the family graveyard just outside, or even within the house itself. The idea was to make the house not just a great larder of local foodstuffs but also a sacred repository of the family spirit. It was animal territoriality writ large, across generations.

The urge to mythologize the estates of the rich in this fashion was deeply rooted. For the Greeks and Romans, ancestors buried around the dwelling place consecrated the soil and forged a sacred link between the land and the people who lived on it. Likewise, prosperous Mayans in the Copán valley kept their loved ones "under their patio floors within earshot of the children and descendants working and playing above them. . . . When the family patriarch stood on the patio and conducted a bloodletting, he knew the ancestors were below his feet—close at hand should he want to call them forth." For the Chinese aristocracy, the physical link with their ancestors enhanced their *tê*, meaning their magical power. Commoners had no great estates, no family names, no ancestral cult, and precious little *tê*. Getting them to forget their names was a key means of keeping them subordinate.

The names of the rich meanwhile were literally enshrined in their houses. At Blenheim Palace, for instance, the dukes and duchesses of Marlborough are buried beneath the floor of the magnificent family chapel. The chapel suffers no graven images of God. A scroll over the organ says "Sing Unto the Lord," and neglects to say which lord. The chapel is dominated by a large, idolatrous statue of John Churchill, depicting him as a Roman military hero rising to the heavens, with Sarah gazing up at him worshipfully and his young sons disporting themselves below, while the figure of history records his achievements with a quill pen, and fame trumpets his glory down through the ages, their combined weight crushing the dragons of envy and spite. The Churchill family pews do not face the altar, but sit opposite this statue at pointblank range. When she was an unhappy duchess here early in the twentieth century, Consuelo Vanderbilt used to sit beneath this statue and entertain disloyal democratic thoughts about "pompous vanity." She also complained about "the smell of putrefaction," a rather potent form of scent-marking. On the other hand, her husband, the ninth duke, breathed it all in deeply. A visiting artist once remarked that the chapel provided nothing to divert one's attention to God. "No," the duke readily agreed, "the Marlboroughs are worshipped here."

This melding of land, dead ancestors, and religion (of a sort), functioned to keep family identity alive. It was a way of reminding the outside world, and one's own descendants, to tread lightly: *This is our land. These are our houses. Be fearful.* "From the song of a certain bird," Konrad Lorenz wrote, "other birds not yet in possession of a territory recognize that in this particular place a male is proclaiming territorial rights. " A very particular male, whose song conveyed his identity, his age, and just how much he ought to be feared. Or as Lorenz put it, the cock doesn't crow to say, "Here is a cock!" It crows to make a far more special announcement, "Here is the cock Balthazar!"

Today we are a bit too democratic to attach any sacredness or magical power to the family name ("charisma" is our preferred euphemism). But keeping a lineage rooted in a family compound still clearly matters for the Kennedys at Hyannisport, the du Ponts in Delaware, the Rockefellers at Pocantico Hills, and many other wealthy families. Such families typically try to bring their far-flung cousinage back to the ancestral estate every year or so to renew their blood compact at a family meeting. As a way of preserving the sense of family over many generations, some of these meetings feature explicitly tribal rituals, including initiation of fourteen-year-olds and celebration of "elders" and "storytellers." (These are, one must add, very wholesome rituals; we are not talking animal sacrifice.) When wealth persists for generations and the family expands to include hundreds of cousins, a family territory—the founding father's camp in the Adirondacks, the island off the coast of Maine, the country house in Scotland or Normandy—may seem like the one real thing holding the family identity together. That and, of course, the money.

Here and there a numinous connection to the sacred family soil also dimly persists. One New England family of my acquaintance has held onto the same tract of land for more than 350 years. In the customary fashion, they have bled and been buried there. A descendant now grows Christmas trees on his portion of the property and a few years ago, another descendant on the property asked if he could buy his Christmas tree live, with the root ball intact. The tree-growing descendant declined, saying he did not want to surrender "the ancestral soil." He was not persuaded even when his cousin offered to dig up some of his own portion of the ancestral soil and cart it over in a wheelbarrow in recompense.

The Teardown

For John Churchill, who managed much larger enterprises with such coolness and control, the construction of Blenheim was to prove an almost endless fiasco. Because he was mostly away at war, he could do little more than dream of his glorious house and acquire paintings and tapestries to furnish it. In marriages of the rich, where the husband is off conquering the world, house-building often falls in this fashion to the neglected wife. It is a recipe for disaster: The wife has too often been denied any practical experience in wielding power, and she is all too aware that her husband's expectations are over the moon, from a lifetime delegating tasks to underlings trained to bark on command. So maybe it was inevitable that Sarah Churchill would commit every sin of every wife who ever built a McMansion. Yet she seems to have done so with her own special gusto, quarreling with the architect, cheating the workmen, placing grandeur above comfort, and wildly overspending their budget. Architects figured it would cost the nation £100,000 to build the Churchills a suitable house. In the end, it cost more like £300,000, about $63 million in today's money.

Blenheim also entailed a teardown, a standard feature in establishing a family estate. People who buy a famous property typically have a choice: They can, on the one hand, assume the mantle of the previous owner. For the producer David Geffen, for instance, moving into movie pioneer Jack L. Warner's old mansion was a way to live out his childhood dream of becoming a Hollywood mogul. The other, more popular option, is to obliterate the name and any other trace of anyone who previously occupied the site, like dogs overmarking one another at the local fire hydrant. For instance, Aaron Spelling and his wife Candy paid $10.2 million for Bing Crosby's 43,000-square-foot mansion in Holmby Hills, then tore it down and put up a 56,000-square-foot monster in its place. ("If you think about it, there are a lot of well-to-do people who have many homes everywhere in the world," said Candy, to explain why two people need so much living space. *"This is all we have."*) Sometimes the houses get put up and torn down without ever actually having been slept in. In Aspen a few years ago, for instance, a restaurant chain entrepreneur bought a new 20,000-square-foot house built by a developer in the Owl Creek neighborhood and immediately replaced it with a more luxurious home. Then he split up with his significant other and sold the new place, which he had also never occupied. Even by these standards, though, what happened at Blenheim was one of the most egregious teardowns in history.

One morning at six I walked out to Blenheim in fog thick as a cloud of cotton wool. It was cold. The young leaves of a chestnut drooped down, soft and tender as the wings of newly emerged damselflies. Doves hooted softly in the branches, and wood pigeons took off with a sudden, toneless pulsing. The palace was just a vague mass in the fog as I passed under its walls and headed down across the great bridge to the far side of the lake. Here, a royal lodge once stood, and possibly a Roman villa before that. English kings back to Ethelred the Unready came to hunt and misbehave. Given the hour and the fog, I thought I might sense their shades still wandering there. Church bells were ringing. The sun was burning a white hole in the sky. The domineering presence of Blenheim was still mercifully hidden by fog. I stood under a massive old oak, with a trunk eight feet across, which might have been a sapling in the 1550s when the future Queen Elizabeth was here as a prisoner of her sister Queen Mary. History in England often feels close like that. But at

Blenheim, it felt as if all the royal ghosts had been swept under a thick carpet of grass, to be forgotten.

The old royal lodge that stood on this spot was "one of the most agreeable objects that the best of landskip painters could invent," according to the Churchills' architect John Vanbrugh. But Sarah Churchill wanted it down. When Vanbrugh suggested an obelisk marking the spring where King Henry II wooed his mistress the "fair Rosamund," Sarah scoffed that if the English put up monuments to "all our kings have done of that sort, the countrey would be stuffed with very odd things." Blenheim was meant to fix the visitor's gaze on just one thing—the eternal glory of the Churchills—and Sarah wasn't about to let former kings obstruct the view.

Throughout the construction, she and Vanbrugh waged a second battle of Blenheim. He had his workmen make a show of demolishing parts of the royal lodge when Sarah came to visit. Then he rebuilt them when she went away again. In the eleventh year of construction, with Blenheim still in her words "a chaos which only God Almighty could finish," Vanbrugh diverted his attention to his great bridge. Sarah had every reason to be furious. The palace itself still didn't have one room fit for her to spend the night. The magnificent bridge also crossed a ridiculous trickle of a stream (later dammed by Capability Brown to create the present lake) and it went nowhere, the main entrance road not yet having been built. Both ends of the bridge stood in midair, short of the steep sides of the glen. Sarah ultimately fired Vanbrugh, had the royal lodge torn down on her own, and used the rubble to fill in around the bridge, in effect building a road to the glory of the Churchills on the bones of dead kings.

The Trophy Case

When a visitor walks through the courtyards of Blenheim Palace today, it's easy to interpret the elaborate ornamentation as just so much Baroque excess or as the theatrical result of employing Vanbrugh, a former playwright, as the architect. But Blenheim was never meant to be merely a great house nor a grand theatrical set. It would be more accurate to think of it as the world's largest trophy case, celebrating John Churchill's many successful hunts on behalf of Queen Anne, and the head on the wall belongs to the French King Louis XIV.

"Look over the archway," my guide remarked, as we stood in front of the palace. "Do you see the statue of a cockerel trying to rise with its wings spread? It's the emblem of France. But the British lion is holding it down and methodically tearing it to pieces." (*Sic transit* cock Balthazar.) And the flaming finials atop the palace towers? "Hand grenades," the guide said. The flames at the top also represent the Duke of Marlborough's coronet, atop a stone sphere symbolizing his power, both bearing down on the three leaves of a fleur-de-lis, symbol of Louis XIV, "upside down in defeat." The guide did the math: "There are four finials on that tower, and there are four towers, so the image of the Duke of Marlborough defeating France is stated 16 times." Like many trophy cases, this one leaves little to the imagination.

Curiously, Blenheim was also an homage to Louis XIV. In 1674, the young Churchill had led an English army under the command of Louis XIV against the Dutch. *Le bel Anglais*, as the French called him then, before he was horribly transmogrified into Malbrouk, was among the first foreigners received by Louis in the *Grand Appartement* at the still unfinished palace of Versailles. That visit and a later stay at Versailles informed Churchill's ideas about splendor. Blenheim was thus perhaps the first of the innumerable imitations of Versailles that have been perpetrated around the planet ever since by rich men wishing to live as grandly as the Sun King—the Linderhof Castle of Ludwig II in Bavaria, the Castle of Eszterhazy in Hungary, the summer palace of the Ch'ing Dynasty in China, on up to the $50 million Versailles begun on the banks of the Potomac by Michael Saylor of Microstrategy, a little before he lost $6 billion in a day during the post-1999 Internet stock market bust.

Blenheim, though, was distinctly an homage to the *vanquished* Sun King. Even now, a pennant bearing the French king's fleur-de-lis hangs from a marble mantle in a state room of the palace, like a scalp. And on the anniversary of the battle each year, a Blenheim functionary climbs into his car and drives down to Windsor Castle to deliver a copy of this pennant to a functionary for the Queen, as the Churchill family's rent for the Blenheim Palace grounds.

If Blenheim was a trophy case, it was a celebration not just of war but also of sexual conquest, a temple to Mars and Venus, chimp and bonobo, alike. The great hallway still has its sixty-five-foot-high ceiling painted with the image of John Churchill as a warrior presenting the fruits of victory to Britannia. It once also contained nine paintings attributed to Titian, "The Loves of the Gods," depicting "a flushed Pluto . . . embracing Proserpine, an avid Apollo grabbing Daphne, an inebriated Bacchus tickling Cupid; all of them very large and pink and sensual," according to the historian Marian Fowler. When William Hazlitt visited in the 1820s, he was moved (possibly in the manner of Strachey after him) by the "purple light of love, crimsoned blushes, looks bathed in rapture, kisses with immortal sweetness," and particularly by a Cupid "who might well turn the world upside down." Fowler writes: "Like Blenheim's martial roofscape, the Titians reflected the raw passion of the age which sired the palace: a crude age of sex and violence in which animal spirits were openly expressed." The great entryway was thus a celebration of "whoring and warring." Unfortunately, a prudish duke in the Victorian era removed the paintings and hid them away in storage, where they were destroyed by fire, along with Rubens's "Rape of Proserpine." Until then they hung as a sort of encouragement to sexual misbehavior—and perpetuation of the lineage—by future generations.

Prospect and Refuge

The houses of the rich are typically situated to command the landscape, and they have been that way since long before Blenheim. Masada, a massive red butte rising up from the Judean Desert, is famous as the site where a handful of Jewish rebels held out to the death against the Roman army in A.D. 66, but it earned an earlier place in history as the archetypal retreat of a rich person. King Herod, a foreigner hated by his subjects, built his palace here more than two thousand years ago and it has survived in remarkable condition. To get to it in the old days, you needed to climb a narrow switch-backing path up a 1,300-foot cliff, at considerable risk of slipping off into the abyss or being encouraged in that direction by skillful bombardment from above. Then you had to pass through a heavily fortified gate into the enclosed cliff top. But you still hadn't gotten within groveling distance of Herod. His home was still further protected, behind a vast fortified storage area, with row after row of jars packed with oil, wine, grains, and other foods. To hold out against siege, and also supply water to Herod's heated bathhouse, huge underground cisterns had been carved out of the rock. The palace itself, three stories deep, was cut into the north face of the cliff, like the bridge on a mighty ship. It commanded a view across the desert to the Dead Sea, and a hedgehog could hardly approach on the road from Jerusalem without the people on Masada knowing about it. Standing on the balcony there, with fan-tailed ravens wheeling around him on the thermals, Herod must have felt, if not quite in command of his fractious citizenry, at least safe from them.

The rich like to make their homes in places like that, on the top of a hill or in a penthouse apartment with a Central Park view or on a bluff above a bay. They also like to have a retreat or bolthole in case the world goes mad. New technology has made it easier to create neo-Masadas off the grid in remote locations. One architect showed me an aerie he had built on a windy peak in the Rocky Mountains, to which his client retreats by helicopter and the propane for the superquiet generators must be trucked in on a dirt road twelve miles long. At another such property, the owner, a mining company executive, has had the heavy beams in the living room ceiling inscribed with the words of the Robert Frost poem: "Two roads diverged in a wood, and I, I took the one less traveled by." The special appeal of such houses is that they combine the sense of being secure, even hidden, with the exhilaration of being able to see for miles in all directions.

The prospect-refuge theory, developed by British geographer Jay Appleton in his 1975 book *The Experience of Landscape*, holds that humans are biologically predisposed to prefer such places, where we have the advantage of being able to "see without being seen." The more expensive the house, the more likely it is to include architectural features that combine prospect and refuge: Balconies, balustrades, bay windows, sun rooms, cupolas, towers, porches, terraces with grape-laden pergolas, trellises, gazebos. It is a leap from Masada to Frank Lloyd Wright's Fallingwater, built in 1939 as the weekend retreat for a Pittsburgh department store owner. But Fallingwater also appeals largely through the combination of nooks and crannies for refuge, with porches cantilevered out over the waterfall for an expansive sense of prospect. Sarah Churchill was working in the same great tradition as a young woman on the make in London, where she made it her business "to observe things very exactly without being much observed myself," and also much later, at the height of her triumph, when the first thing she asked for at Blenheim was a comfortable room with a bow window.

Standing in such a place, according to Appleton's theory, is like being in a forest at the edge of a clearing. More to the point, it is like being hidden at the edge of a savanna 100,000 years ago, when our ancestors were still hunter-gatherers in Africa. I have camped in such places, most recently on a trip that took me first to Blenheim Palace and then onward to the Okavango Delta in northern Botswana, and I have experienced the widely reported sense of being at home in such habitats. (Botswana, I mean; Blenheim a little less so.) The stars were spangled across the sky in smoky clusters of light, and I lay in my tent listening to the distant rumble of lions and the doleful keening of jackals. In the morning, I stood hidden in a cluster of trees overlooking a flood plain and watched antelope grazing on the open plain, and it felt good. Is it plausible, though, to think that such an archaic setting could still influence the way modern humans with money build their homes?

The Savanna Hypothesis

It's difficult to demonstrate, in a scientifically persuasive way, that humans display a positive bias in favor of certain habitat cues. A negative bias leftover from our evolutionary past is, on the other hand, relatively easy to demonstrate: Deadly snakes or spiders have never been a part of daily life for most urbanites; in temperate regions, it may be hundreds, or even thousands, of years since they last posed a real threat to our ancestors. But the visceral fear stays with us. Scientists have conducted Pavlovian conditioning experiments in which volunteers are repeatedly exposed to threatening images. Fear of everyday modern hazards like handguns and frayed wires quickly fades away. But fear of snakes and spiders, as measured by heart rate and other autonomic nervous system activity, persists long afterward. It's in our genes.

Tantalizing evidence suggests we also have a biologically prepared *positive* response to nature. Texas A & M researcher Roger Ulrich, for instance, has shown that people who view a calming nature video after a stressful experience have markedly lower muscle tension, pulse rate, and skin conductance activity after just five to seven minutes. This translates into significant medical benefits. Ulrich monitored patients after gall bladder surgery and found that those assigned to a room looking out on trees needed far fewer painkillers than patients in rooms looking out on a brick wall. Open-heart patients in rooms with nature scenes on the wall had lower blood pressure and smoother recoveries than patients with blank walls or abstract art.

Surveys and experiments by biologists indicate that people in different cultures respond positively to the same elements in a landscape: Open grassland, scattered stands of branchy trees, water, changes in elevation, winding trails, and brightly lit clearings, preferably partly obscured by foliage in the foreground. It's a landscape which invites exploration, promising resources and refuge at the same time. The changes in elevation—a view of distant mountains, for instance—provide a landmark to help the viewer orient himself in the scene, much as do the rocky outcrops which stud the vast open expanse of the Serengeti.

The growing evidence for an evolutionary influence on our ideas about habitat encouraged University of Washington biologist Gordon Orians to propose a sequel to the prospect-refuge theory. He called it "the savanna hypothesis." It holds that habitat cues that were a matter of life and death when Lucy was dragging her hairy knuckles around Olduvai Gorge survive to the present day as the genetic basis for our aesthetic sensibilities—for instance, in Martha Stewart delicately pruning roses at her house on the coast of Maine.

The best possible prospect, according to the savanna hypothesis, includes the promise of food for dinner: Not just a water hole, but large mammals grazing scenically around it. To separate the aristocracy from their handsome livestock, British landscape designers developed the ingenious device called the ha-ha, basically an impassable trench. It allowed the animals to graze on fields that appeared to be as open and unobstructed as the Serengeti itself. Proponents of the savannah hypothesis argue that we value flowers in the foreground not just for their beauty, but also because they promise fruit and honey. Anyone who has ever contemplated the magnificent centerpiece at a charity ball, or in the pages of *Town and Country*, will know that we prefer flowers to be big and asymmetrical, traits that correlate with greater nectar content. Flowers and herds of grazing animals are literally good medicine. Like bare willow stems for yellow warblers, they soothe us with the promise of prosperous times ahead.

Orians has gone so far as to argue that the celebrated landscape of the English country house, that quintessence of the rich man's estate, was an unconscious attempt to re-create the environment of the African savanna. Moreover, he has produced intriguing evidence of "savannafying" tendencies associated with this landscape. For instance, Orians and coauthor Judith Heerwagen compared sketches made on the spot with the finished works produced later in the studio by eighteenth-century landscape painter John Constable. They found that the artist consistently "savannafied" reality, opening views to the horizon, making water features more conspicuous, adding large grazing mammals, and stripping away foliage to expose tree branches. That is, he enhanced habitat cues most likely to gratify the inner primate. The proper aesthetic response to a Constable landscape may thus be to stand and softly hoot.

Like Constable, landscape designer Humphry Repton also made "before" and "after" drawings of the English countryside. But Repton actually bound both sets of images, with explanatory text, between red leather covers and presented these "Red Books" to wealthy landowners. The aim, often successful, was to persuade them to hire him to make the proposed changes to their estates. Repton was a follower of the great eighteenth-century landscape designers William Kent and Capability Brown, who popularized the romantic "naturalized" style on English country estates. It was a movement away from the rigid geometries of classical gardening toward an idealized curvilinear wilderness. When Orians and Heerwagen studied eighteen before-and-after sets, they found that Repton, like Constable, had removed trees to open views to the horizon, enhanced water features, and added no fewer than two hundred grazing mammals, though they were presumably irrelevant to his actual design. On the apparent premise that rich people should not have to look at riffraff, Repton also gentrified the people on the estates, in one case replacing a peasant wielding a pickax with an artist painting the newly picturesque view.

Woburn Abbey, home of the dukes of Bedford, was one of the country estates Repton actually landscaped. So I phoned up the current chatelaine, Lady Tavistock, and tried out Orians' idea that Repton had created, as it were, a little piece of Tanzania there. "*No!*" she replied, in a horrified tone. "The idea was to bring out the natural character of the English countryside." But to achieve a "natural" effect at Woburn, Repton had merely employed his customary means of cutting down forests to create open vistas, introducing water holes, and otherwise turning the grounds into a savanna. Since the 1950s, part of the Woburn Abbey estate has also functioned as a safari park, featuring toothsome African animals from addax to zebra. Possibly Lady Tavistock was thinking of the natural character of the English countryside as it existed 25,000 years ago, when elephants and rhinos still wandered there?

The savannifying tendencies of the British aristocracy date back at least to Norman times. William The Conqueror's son Henry I, who reigned from 1100 to 1135, introduced lions, leopards, and other exotic animals to his walled hunting grounds in Woodstock, at what is now Blenheim Palace. In the 1760s, Blenheim underwent the naturalizing movement at the hands of the master, Capability Brown. By damming the River Glyme, Brown created two large, serpentine lakes, cinched across the middle by Vanbrugh's magnificent bridge, an improvement hailed by everyone who has ever visited Blenheim, including Brown himself. "I think I have made the River Thames blush today," he remarked. Brown also converted the rest of the park to a sublime landscape of cultivated wildness, all grassy fields and sinuous forest edge. His work survives to the present day, as well as in plans for landscaping well into the future. Sheep still graze on his vast lawns and, at regular intervals, his ornamental beech trees stand ready with their low, commodious branches spread out like an eternal refuge, lest any Churchill become spooked by the busloads of tourists passing through.

Bonobo Gone Bad

The political climate in which John and Sarah Churchill lived was astonishingly venomous and, as the war and the construction of Blenheim both dragged on, the Churchills often got the worst of it. Not content to ridicule Churchill for his well-known love of money, for instance, one critic accused him of sending his officers to certain death so he could profit from reselling their commissions. (The critic backed down when the otherwise unflappable Churchill rightly deemed this grounds for a duel.) Even in this harsh context, Sarah's own rhetoric stands out for sheer rancor. She quarreled with almost everyone and became increasingly dictatorial in her efforts to force the Churchills' own political inclinations on Queen Anne. When the Queen balked, and ultimately came to depend on another of her attendants, Sarah circulated the rumor that her rival was "a slut" and a "dirty chambermaid," who had won the Queen's heart through "the Conduct and the Care of some Dark Deeds at Night." She actually sang a song to this effect, to the tune, oddly, of "Fair Rosamund."

Not surprisingly, the Churchills fell out of favor. In 1710, their political adversaries came to power. Funding for Blenheim Palace was cut off, and the unpaid workers were left to starve. The Churchills eventually went into exile, to return only when George I succeeded Queen Anne in 1714. They did not actually move into Blenheim Palace, which they completed partly with their own funds, until 1719, and John lived to spend only part of two summers there. By then, he was a stroke-addled old man. Sarah lived another twenty-two years after his death, the exact moment of which is carved in stone on one of the towers. But she seldom came back.

Who's He When He's at Home?

For people of ordinary means, one of the great puzzles about wealth is why it compels the rich to have such large houses, and so many of them. They rarely inhabit these houses, rather like Kurt Vonnegut's fictitious

millionaire Winston Niles Rumfoord, who had a mansion to rival in substance the Great Pyramid of Khufu but somehow got himself stuck orbit-
ing through the cosmos, so that "the quondam master of the house,
except for one hour in every fifty-nine days, was no more substantial than
a moonbeam."

Though the rich seldom intend it, their houses often serve as brief
stops on a journey away from the real world. The process of removal usually starts with simple convenience. Ducking the routine hassles of ordinary life is, after all, one reason people want to have money in the first
place. The Hotel Bel-Air in Los Angeles, for example, frequently hosts a
couple who have their luggage, all ten Louis Vuitton bags, packed and
FedExed both ways. Even if circumstances oblige them to fly commercial, they need never endure panty-handling security guards and other
indignities of the airport baggage experience. It's expensive. But if we
could afford it, who among us would schlep?

"This removal process is very seductive and very addictive and it
tends to feed on itself," says Peter White of Citibank. "The more you get
removed from people, the more you react in ways that remove you from
people. When I was a lawyer, I got to the point where I couldn't conceive
of riding in coach. I have a friend who can't conceive of riding in a commercial carrier. You get used to things that put you into a mode where
your interaction with other people is dramatically reduced, and you
become a world unto yourself."

Retreating to large, guarded homes behind closed gates within exclusive communities may be merely prudent: White has had three clients
who have suffered kidnappings in their families and one, in Latin America, who was killed. Even if physical danger is unlikely, being rich also
puts people in a discomfiting state of status tension with their financial
inferiors. It may not show up as anything more obvious than momentarily lifted eyebrows, or an expression of bafflement giving way to fleeting
contempt. Or it may be ludicrously overt: One unfortunate woman
recalled a leftist boyfriend with a peculiar idea of what it means to mount
the barricades: "He felt when he was screwing me, he was screwing the ruling class." Becoming rich accustoms a person to flattery and obedience
within his own world, and this can make the insolence of the world at large
insufferable. At one small airport that routinely serves the rich, a passenger

was indignant at being delayed by bad weather. When he had rapidly run through other avenues of complaint, with mounting fury, he finally demanded, *"Do you know who I am?"* The clerk picked up the public address microphone and announced, "The gentleman at the counter does not know who he is. If anyone can help him, would you please come forward?"

In her study *The Experience of Inherited Wealth,* psychologist and wealth counselor Joanie Bronfman quotes one rich woman: "I see now why rich people isolate themselves from the rest of the human race. Why they go off to clubs and live in fancy communities. . . . I think it's because rich people culturally and protectively have realized that it's scary out there. People don't like you, they resent you. It's not only that rich people are snobbish people, it's also that they feel misunderstood and discriminated against in the outside world."

Hence the rich do much of their traveling among a handful of familiar destinations, like Lyford Cay, Palm Beach, St. Moritz, and Majorca. They want the sense of security and control that comes from being among people they know, and people who know them. One prominent socialite travels by private jet from her home in Texas, and when she visits one of her other homes, she expects the maid to be waiting, in uniform, at the foot of the aircraft stairway, along with her car and driver. There is very little chance, between one door and the other, that she will meet anyone who does not know who she is. (A valet notes that she never smiles. This is apparently reserved for the event photographer.) A billionaire in his fifties has a different mistress at each of his three homes. (Of his wife, a friend says, "I don't think she gives a damn. She has an agenda, a political agenda, and he can make it happen.") The rich like to own houses all along their normal migratory circuit to be, as much as possible, always within the comfort of their own territory.

Their homes become the world, almost literally. A few years ago, at a New Yorker's weekend retreat in a gated community in Connecticut, I opened an unprepossessing doorway off the front hall, descended a long stairway, and found myself suddenly in Times Square circa 1928: There was a theater marquee ablaze with light and a box office with 24-carat-gold-leaf pilasters, staffed by Edna the ticket-taker, a hand-crafted mannequin. I opened the glass doors to the lobby, their brass push plates embossed with the old italicized Paramount Pictures logo, salvaged from one of the original 1920s movie palaces. Beyond the refreshment area with its hidden popcorn machine and its posters for *Please Don't Eat the Daisies* and *Lord of the Jungle,* was a foyer with a carved mahogany fireplace mantel and armchairs upholstered in Scalamandre. Beyond that yet another anteroom, and then the theater itself, a Beaux Arts extravaganza, with a twenty-foot-high coffered ceiling, fluted pilasters, and tapestry side panels. Footlights bathed the tassels of the stage curtain. The house lights dimmed, the curtain swayed open, and the show began. Now and then, I noticed the reflected light of the film flickering softly off the gold leaf of the proscenium frame. There were four twelve-foot-long Bose subwoofers suspended underneath the orchestra floor. The sound quality was such that you could watch *Apollo 13* and feel as if you were hanging onto a tailfin at liftoff. Unfortunately, we were watching *Casper,* about a friendly but insipid ghost.

The whole theater had cost upwards of $800,000, and at least in theory the aim was to achieve the opposite of the couch-potato experience. The scale and literal theatricality of the place were meant to encourage the owner to organize little movie parties for friends. *Lawrence of Arabia* with babaganoush at the intermission, yes; *Silence of the Lambs* with fava beans and a nice chianti, perhaps not. "When you're in that theater, you're in another time and another place. It's a sanctuary," the owner told me. "And it's a window on the world." But what stuck with me was something else he said about the theater, which is equipped with every electronic game imaginable and a joy stick fit for the gods: "The whole idea was never to have to go out, and I wanted my kids never to have to go out."

Since then, as the outside world has come to seem increasingly unfriendly, the idea of the home theater has evolved into new homes with entire entertainment wings built in. I visited one such home under construction atop an oceanfront cliff south of Los Angeles. The basement included a 6,000-square-foot garage for the owner's car collection plus his own mock dealership, Frankie's Motor Works. Also downstairs, Kay's Jewelers (with an 8,000-pound door on the safe), the Blossom Daisy Cafe (a working diner), the Rialto Theater, and Beach Lanes Bowling Alley formed a private streetscape. Upstairs, the house looked out on a spectacular stretch of cliff angling out to the west. The cliff face actually reappeared inside the house, in a huge atrium. But the inner cliff face turned out to be a fiberglass false-front, concealing a spiraling water slide down to the swimming pool. In the same neighborhood, another homeowner was in the final stages of planning a home entertainment complex built around a Tuscan village square. In addition to the usual fare (a theater based on the Paris Opera, a pizzeria with a wood-burning brick oven, an Olympic-length lap pool), he was going to have his own nightclub and Roman baths, with *frigidarium*, *tepidarium*, and *calidarium*. The natural question, which even the owner asked as he rolled out his blueprints, is whether any family can live on such a scale and in such isolation from everyday life and yet remain normal.

Inspecting the Privates

"Are the privates open today?" a guide at Blenheim Palace asked a colleague one afternoon during my visit, using the tourist industry's oddly off-color jargon for the Churchill family's private apartments. Visitors can tour them for an additional fee, depending on whether the duke is in residence. "We're in a gray area where, although the duke is in residence, he's not actually here," the colleague explained. So the privates were open.

"Because this is a private residence," the guide began, as we stood in a long vaulted hallway, "there are lots of personal knickknacks around. Love them, if you like. Hate them if you like. But no fingerprints." Under glass, a guest book lay open to June 27/28, 1936, when the country house weekend was still a fixture of aristocratic life. The left page was given over to a single signature, Edward R.I. (*Rex Imperator*), and the right to a dozen or so other guests, among them Wallis Simpson and her husband Ernest, along with Winston and Clementine Churchill. Serious company.

The privates were of course also serious living space. But like the rest of the palace, they felt oddly impersonal. The books in the sitting room were the usual coffee table offerings, *Edwardian Portraits*, *The Fauves*, *Impressionism*, each exactly where it had been when I'd visited a year or two before. On her bed, the duchess had some pillows decorated by a daughter. Also precisely as before. It seemed as if the owners themselves had to be careful not to leave fingerprints on the objects in their own display case.

But John and Sarah Churchill still had their fingerprints everywhere. Among the videos stacked beside the television in the smoking room was one titled *History of the Duke of Marlborough.* Another, *France vs. England,* was either a soccer match or a war. In the family's private dining room, the chief feature was a series of three tapestries called "The Art of War," given to the first duke by the people of Brussels. Family members at breakfast could look up in one direction and see an eighteenth-century soldier felled from his horse, blood gushing from his shattered skull. Or they could look the other way. "On top here," the guide said, indicating a painting, "is the Duke of Marlborough, looking down on everyone as usual." I began to feel almost sorry for them.

Not too surprisingly, the inhabitants of Blenheim have generally proved themselves to be misfits and scoundrels. If, as one historian conjectured, Blenheim has often been "an unpleasant place lived in by unpleasant people," the unpleasantness may well result from what the house itself has done to the generations condemned to live in it. The fourth duke of Marlborough was the first to treat Blenheim as a residence, beginning in 1762, and his large family appears to have been happy there, at least in the early years. But the duke later became a hypochondriac and a recluse, wandering the vast rooms at Blenheim in silence. He once went three years without uttering a word, a silence broken only upon being informed that a French author was coming to visit, at which point he very sensibly roared "take me away." The fifth duke was a spendthrift and virtual bankrupt, the sixth used a sham marriage to seduce and impregnate an innocent girl (which was all very well, except that he got caught at it), the seventh, described by A. L. Rowse as "a complete full-blown Victorian prig," sold off the family treasures, and the eighth married a fat American with a mustache and a $5 million fortune. (He made her change her name from Lillian to Lily lest the unruly press make unfortunate rhymes with "million." Despite their marriage, he also kept a nude portrait of his mistress on his bedroom wall.) In the 1880s, the British prime minister William Gladstone put it comprehensively: "There never was a Churchill from John of Marlborough down that had either morals or principles."

Apart from the burden of the founding father's mythology, the appalling expense of maintaining his monument also kept his descendants imprisoned and incapacitated. What Beatrice and Sidney Webb said about another great country estate, Luton Hoo, was even more true of Blenheim; it was "a machine for the futile expenditure of wealth." To keep the roof overhead, the Churchills were already charging admission to the general public in the 1780s, and nineteenth-century visitors could fish in the lake or shoot game for an hourly fee. In the 1960s, when the tenth duke was still trying to maintain the palace as a tourist attraction and a family residence at the same time, a guest complained that the duke was unwittingly "turning him and his friends into exhibits, like freaks or animals to be gazed at. I felt this most acutely when playing croquet in the public eye, surrounded by a chain to keep the not unnaturally inquisitive visitors off the lawn. I thought at any moment they would throw us stale bread or nuts as in the zoo." The Churchills, unlike their guests, were already old hands at living in a cage.

Bird in a Gilded Cage

The real problem may simply have been that the cage was so big. If you listen to the rich talking about the quality of their home lives today, the word you hear over and over is "isolation," a sense of being not merely separate, but alone. The withdrawal from contact with other people, which begins as a quest for convenience or security, often ends up being borderline pathological: The Hotel Bel-Air has a regular guest who insists that his room have a new mattress every time he visits, that it be wrapped in eight sheets, also new, and covered with a new blanket wrapped in two sheets. He also gets eighteen bath towels, eighteen hand towels, and six bars of soap. New, new, and, of course, new. A woman planning a 30,000-square-foot house in Aspen has included two swimming pools, one for her children, and another, uncontaminated by human touch, for herself.

Yet we are built for human contact. Other social primates may wander the landscape by day, but they often re-group to feed together and reinforce their social bonds with mutual grooming. By night, they huddle together in the safety of their trees. Put each of them in a different tree, over an area the size of a football field, and you end up with a lot of panicky, insecure monkeys. Likewise rich humans in their mansions.

"The ordinary child sleeps in a room next door to his parents or nearby," a woman who grew up with inherited wealth, in a big house, told psychologist Joanie Bronfman. "He can practically hear his parents breathing through the walls. But I remember being locked in my room for some reason, probably because I used to try to get out of my room and go and visit my brothers. I was apparently a bit of a nuisance. But I really needed my parents. I really wanted them. I don't remember what, but it was for some urgent thing, and I called and I called and no answer. I cried and no answer, and finally I just lost control and I had a tantrum. I threw my piggy bank at the door until the door was all bruised and splintered. I must have gone like that for a half hour and thinking that I'm sure people are going to hear me and come in and punish me. At least I would get attention, at least they would discover me. Nobody came and nobody came and I finally came to the end of the tantrum. . . . When I got the strength together, I crept out through my window and wandered downstairs. My mother was having cocktails with my father and some guests and she had no idea anything was going wrong. She said, 'Hello, what brings you down here?' It just made me feel terrible. I had absolutely pulled out everything in my arsenal and they didn't even know that I had any trouble at all."

Another of Bronfman's interview subjects grew up in a mansion atop a hill: "The people in the town weren't suitable for me to play with. There were no houses closer than two miles away. They wouldn't let me ride my bicycle off the hill. . . . Eventually, Mother died, Father got sick, and I got my money." He used it to buy himself his own island, off the coast of Maine, which he described as "a gorgeous thing, it's like religion when I go out." It only occurred to him afterward that an island "is a perfect symbol for isolation."

At Blenheim, the most poignant tale of being adrift on the sea of interior space came from Consuelo Vanderbilt. She was an eighteen-year-old American in 1896, when her ambitious mother forced her to marry the ninth duke of Marlborough for his title. Vanderbilt thus became the most prominent of the "dollar duchesses," an armada of rich American brides imported to rescue the fading aristocracy. (The duke's uncle Randolph had also married a wealthy American, Jennie Jerome, and at least as long as the duke was unmarried and childless, their son Winston was nominally in line for the title.) Vanderbilt brought the Churchills a dowry of $2.5 million in Beech Creek Railway Company stock, which the duke desperately wanted to restore Blenheim to its former glory. The roof "will now receive some much needed repairs," the *Washington Post* dryly commented, "and the family will be able to go back to three meals a day." Soon after their marriage, Consuelo's husband informed her that her role was to be "a link in the chain." His grandmother admonished her to produce a male heir promptly, "because it would be intolerable to have that little upstart Winston become duke."

As a Vanderbilt, Consuelo was entirely accustomed to living on the gargantuan scale, but the discomforts of Blenheim were new to her. "It is strange that in so great a house there should not be one really livable room," she complained later. Nor was she much charmed by waking up each morning to face a marble mantel in her bedroom inscribed by her late father-in-law with this *billet doux* to his own duchess: "Dust Ashes Nothing." But the real chill in her new life came from her husband, the duke, who cared about little other than the glory of Blenheim and the Churchill lineage.

As the early years of their marriage passed and the chill set deeper, Vanderbilt put the great John Churchill to practical use. One of the treasures of the palace was a 100-pound silver centerpiece depicting the first duke on horseback, bending down to write a dispatch informing Sarah of his victory at Blenheim. Vanderbilt used this centerpiece as a *cache-mari* or "hide-the-husband," positioning it so she would not have to watch her own lugubrious duke fiddle endlessly with his dinner. "As a rule neither of us spoke a word," Vanderbilt wrote. "I took to knitting in desperation and the butler read detective stories in the hall." When she decorated their house in London, Vanderbilt placed an image of John Churchill at one end of the hall, and at the other, her own ancestor, Commodore Vanderbilt, so the two of them could glower at one another like rival pirates. The marriage ended in divorce.

And Yet...

The lives of the rich, their chronicles of solitude and plush suffering, offer ample testimony against the big house as a way of life. Yet, from a Darwinian point of view, the big house often performs exactly as intended. It keeps family identity alive and it assists, none too subtly, in the propagation, legitimately or otherwise, of the family's genetic heritage. The big house has the power to bend the most subversive in-laws—or "out-laws," as they are often known in rich families—into the larger mythology. Even Vanderbilt did her duty, producing two sons, or as she put it, "an heir and a spare." Long after her divorce from the duke, she also chose to be buried with the family's "smaller fry," including Winston Churchill, in the churchyard of Bladon across the park from the house.

Moreover, John Churchill's descendants still sometimes actually live in his monument, three centuries after his celebrated victory. (It reminded me of a tribal chieftain's house I once visited in Uganda, where his widows faithfully tended his altar years after his death.) Blenheim still does not contain a single really livable room. So even when he is officially in residence, the eleventh duke of Marlborough typically goes home to a more private estate a few minutes from the palace. But he also sometimes hosts visiting friends, dignitaries, and shooting parties (who pay handsomely for the privilege) in Blenheim itself.

The duke turned out, when I eventually met him on another trip to Blenheim, to be a tall, thin fellow in his 70s, genial, with slightly protuberant blue eyes and wavy gray hair combed neatly back. He was dressed in a hand-sewn, double-breasted suit, with the gold chain of a cigar-cutter looped through the buttonhole on his lapel. His fondest memories of his childhood in the palace were of riding around the basement, or undercroft, on his bicycle. He also recalled that his father, the tenth duke, used to drive around the garden in an old sports car while the kids chased him on bikes and tried to pelt him with tennis balls—possibly the single most appealing thing I have heard about any of the dukes in their entire history. He admitted that he had not actually visited all 187 rooms in the palace, merely "the majority of them," and that he felt no strong connection to John Churchill. But he was deeply proud to have known his cousin Winston.

At Blenheim, Winston had achieved the miraculous, managing not just to step out of the oppressive shadow of his great ancestor, nor merely to turn himself into one of the great figures of the twentieth century, but also to remake Blenheim as his own shrine. With the Churchill family's preternatural knack for tending the fires of its own glory, he had arranged to get himself born at Blenheim, two months prematurely, when his parents were making a family visit. Moreover, he had had the good sense to be born not in some awkward bedroom upstairs in the private apartment, but in a suite just off the main entry to the palace, where his birthplace could conveniently become a national monument.

In an exhibit area there, Winston's voice now plays over and over, weary and implacable, guiding his fellow citizens through the darkest hours of World War II: "We shall defend our island whatever the cost may be. We shall fight on the beaches. We shall fight on the landing grounds. We shall fight in the fields and in the streets. We shall fight in the hills. We shall never surrender."

A visitor to Blenheim naturally wonders, as Winston's own son wrote, "whether ancestry or environment" should get the credit for his heroic achievements. Winston was nine generations removed from John Churchill, who thus contributed at best a tiny fraction of his genetic heritage, and at worst (in the very likely event that any link in the chain proved unfaithful) nothing at all. But it would be hard to exaggerate the influence of Blenheim, with all its pomp and privilege.

Winston was baptized in the Blenheim chapel, cherished his childhood visits, and came to revere the great duke. As a young man, he wrote a four-volume biography intended to restore John Churchill to his rightful place, with Wellington and Nelson, among England's greatest heroes. A critic of Winston's *Marlborough: His Life and Times* complained that he often seemed to equate himself with his ancestor. Having spent two formative years immersed in the first duke's papers in the Blenheim archives, Winston often wrote to his future wife Clementine in the same words with which John had addressed Sarah: "My dearest soul." He also used the palace as bait to lure Clementine into marriage. "I think you will be amused by Blenheim," he wrote, badly stretching the limits of upper-class understatement. The big house, with its land, its titles, and its rich mythology served as an effective courtship tool, as it had with Jennie

Jerome, Consuelo Vanderbilt, and countless others named and unnamed before them. Winston proposed marriage when he and Clementine were seated on a bench in a temple on the palace grounds, and they passed their honeymoon at Blenheim. At the same time, Winston never suffered the crushing burden of Blenheim. He drew inspiration from it without ever having to pay the price of its upkeep.

On the tinny newsreel soundtrack in the Winston Churchill exhibit, triumphal brass alternates with plangent strings. But the voice itself is seldom emotional, never shrill, always reasoned. Its slow, deliberate, repetitious cadences resonate with the unshakable power of Blenheim and with the achievements of John Churchill, who also once liberated Europe: "You ask what is our aim. I can answer in one word. Victory. Victory at all costs. Victory in spite of all terror. Victory however long and hard the road may be. For without victory, there is no survival."

Winston Churchill recognized that the building of Blenheim on a scale to rival the pyramids ultimately cost John Churchill his place among his nation's heroes. "How much better it would have been had he been cut off in his brilliant prime, a cannonball at Malplaquet," he once remarked, with a ruthless eye for family reputation. "Indeed, his happiness lost much, and his fame gained nothing, by the building of Blenheim," Winston wrote. "However, Blenheim stands, and Marlborough would probably regard it as having fulfilled its purpose if he returned to earth at this day."

Winston would have been too politic to say it out loud. But at least in retrospect, the subtext seems obvious: If the big houses of the rich and powerful really do in the end serve some worthy purpose, Winston Churchill—upstart, glowworm, hero—was himself the living proof of it.

10

The Temptation of
Midnight Feasts

Sex, Anyone?

Whereas their husbands were enormous blundering tanks of animals, the wives were slim, sinuous and sexy, with their neat pointed faces and big melting eyes. They were the personification of femininity, graceful to a degree, beautiful, coquettish and at the same time loving. They were heavenly creatures, and I decided that should I ever have the chance of being an animal in this world I would choose to be a fur seal so that I might enjoy having such a wonderful wife.

—GERALD DURRELL

WHEN EVOLUTIONARY PSYCHOLOGISTS TALK ABOUT HUMAN SEXUAL behavior, they tend to draw analogies from the animal world and they particularly like to talk about hangingflies. These inch-long predators live by the thousands in the temperate forests of North America. They specialize in catching other insects, injecting digestive enzymes into them, and sucking out their innards. So the analogy to the behavior of rich people may seem remote. But when a male hangingfly wants romance, he goes out and catches an even bigger insect than usual and advertises his catch to the female world. Male and female pair off in the undergrowth, hanging by their forelimbs face-to-face like trapeze artists about to attempt an aerial minuet. He clutches the dead insect in his hind legs and holds it up to her as a nuptial gift—or to put it in human terms, he buys her dinner and she allows sex to follow.

Neither male nor female is a patsy in this partnership. If the dinner is too small, she throws him out before he can do much good. It's a variation on the "diamonds are a girl's best friend" theme, and bigger diamonds, or dead insects, make better friends. It takes twenty minutes of vigorous copulation to get her to lose interest in other males and lay her eggs—and he gets twenty minutes only if he brings her a big gift. On the other hand, when his twenty minutes are up, the male may grab back his nuptial gift and fly away with it to seduce other females.

Does this begin to sound terribly familiar? You are probably already thinking of offensive analogies. For instance, when Donald Trump's prenuptial agreement offered second wife Marla a piece of the real money after five years of marriage, and he then dumped her in year four, was he not managing his reproductive assets in a manner and with a timeliness worthy of a hangingfly?

The leap from hangingflies to humans is of course perilous. In our species, both males and females typically consider the resources a potential long-term mate can bring to the relationship. Upper-class men of past generations weren't merely seeking good bloodlines when they did their dating mainly out of the pages of *The Social Register* or *Debrett's*, nor when they asked one another that odd question, "What does her father do?" "I won't say my previous husbands thought only of my money," the Woolworth department store heiress Barbara Hutton once remarked, "but it had a certain fascination for them." Yet in study after study, human females demonstrate a far more pronounced inclination to seek a partner with resources. Contrary to feminist expectations, women with good prospects of their own tend to place even greater emphasis on a man's financial status. Or as a Dallas commercial real estate saleswoman put it, commenting on rich men in general: "They have a little caption over their head that says, 'Let's go!'" Feminists believe the urge to "marry up" the social ladder is cultural, a by-product of economic discrimination and the blighted sensibilities of commercial real estate saleswomen. Evolutionary psychologists say it's also biological: Women make a huge parental investment in their offspring, so in our evolutionary past there was a significant survival advantage in finding a helpmate who could provide compensatory effort and resources.

The evidence for the parental investment argument in other species is strong. Where females bear most of the cost of rearing offspring, they are exceedingly scrupulous about choosing males who demonstrate greater fitness or, like the hangingfly, provide valuable resources. Insisting that the male put some food on her plate isn't the animal equivalent of prostitution, a simple quid pro quo food-for-sex exchange. It's a way for the female to get the energy she needs to reproduce, and also a way to judge the merits of a prospective mate. In common terns, for instance, courtship feeding of his mate is a reliable indicator of a male's subsequent parental feeding of their young.

In humans, biology puts the cost of reproduction almost entirely on females. The father's contribution equals the mother's in just one regard, which happens to be the payoff: He provides half the child's genome. Otherwise, the disparity in effort is appalling. A woman produces just four hundred eggs in a lifetime, a man four *trillion* individual sperm. This works out to about 125 million a day, or to carry it to a ridiculous extreme, which is after all the male way, a little more than 86,805 potential human beings a minute. His opportunity cost—what he gives up in choosing one mate over another—is thus almost nil. And her cost is almost infinite. On top of that, the woman must then invest 80,000 calories in her pregnancy, roughly the energy it would take her to run from New York to Chicago, plus another 182,000 calories to nurse the baby for a year, almost enough energy to plug onward to San Francisco. Or bust. Meanwhile, the male's direct contribution from his single intrepid sperm is about .000000007 of one calorie, or not quite enough energy to roll over in bed and fart loudly. Is it therefore any wonder that women like men who demonstrate an ability to help out? Or that somewhere in our benighted past, male earning power became a useful gauge of reproductive potential, much as men still gauge female reproductive potential primarily on the basis of pretty faces or nice money-makers, well-shaken?

Evolutionary psychologists are careful to note that they are merely trying to explain the mechanisms that underlie our everyday behavior, not endorse them. Moreover, these mechanisms are not mandatory. Evolution need not dictate our romantic or economic choices any more than it forces us to gorge on fats and sugars. We are talking merely about *propensities*, which culture enables us to transform or reject. A rich woman can marry a taxi driver, a prince can choose a fast-food counter-girl with an unfortunate waist-to-hip ratio, and everyone can live happily ever after. But if it doesn't usually happen that way, it may at least be good to know what we are up against.

A Crown of Money

Early in 1976, a beautiful young woman of Lebanese Christian back-ground—we will call her Sonia—was working as a waitress in a restau-rant in Paris, when a Middle Eastern man, twenty years her senior, with a thick mustache and an elegant air, took notice. Amir—the name is also fictitious, as he has since become one of the richest men in the world—kept coming back. "One evening he made me a crown out of money and sent it to me. All 500 franc bills. A crown with a tail. A long tail. And I didn't understand the gesture." Another waitress of more refined under-standing counted thirty bills in the tail alone and figured the crown was worth at least 15,000 francs—about $3,000 at a time when Sonia's entire monthly food budget was $60. "I had to wear it and come and say 'thank you.' That was the order of the manager."

Amir told Sonia the money was a gift. Then he added that she should use some of it to do something about her hair. "So I slapped him in front of everyone and I didn't take the money. But then he wouldn't let go. He kept calling every day." When her birthday came around the next week, he invited her friends to a feast with caviar and foie gras. "And I did like foie gras." He told her that a pretty girl like her shouldn't be working in a restaurant and that he would take care of her. He said he "wanted to pro-tect me, love me, show me the joys of existence. Easterners know how to say these things without sounding ridiculous." A few weeks later he sent her a letter, with a ticket attached, inviting her to meet him for a weekend in London. When she declined, he said, "Either you come or you'll never see my face again." Sonia went and in due course, against the warning of her siblings that such a man would inevitably prove unfaith-ful, she married him. "A girl like me . . . couldn't resist for too long such a romantic man."

The Great Amalgamator

This recipe for romance is a well-established fact of life among the rich: Take a man who is arrogant, unpleasant, or even ugly, add wealth, and the result, as if by alchemy, is irresistible. Money is the great aphrodisiac, or at least the great amalgamator. The more cultivated sort of rich man may well be horrified by the idea that a woman would want him for his money, particularly if he has merely inherited it. Yet in pursuit of a desirable woman he will at some point almost inevitably make a display of his wealth, like Winston Churchill showing Blenheim to Clementine. The normal practice is to embellish his wealth somehow, to do something worthy with it and bring an element of individuality into the equation. But it isn't strictly speaking necessary. In 1938, for instance, J. Seward Johnson courted his future wife with a gracelessness that makes a crown of 500 franc notes seem positively romantic: "He kept proposing to me all the time," she later recalled. "One day he came with some papers, they were Johnson & Johnson lists of figures. I wasn't interested in figures, but I remember that one of the figures was $92,000, which he told me was his quarterly allowance from his stockholdings. Seward said, 'Lucinda, you know I do nothing constructive with my money. You have friends, you have causes, you could do something with my money.' I thought about that, and the next time he proposed, I accepted."

The power of their wealth over women (and incidentally over just about everybody else) is often the main reason men want money in the first place. Aristotle Onassis had a knack for stating things baldly, as one might expect of a man whose barstools were upholstered with sperm whale scrotums. He once said, "If women didn't exist, all the money in the world would have no meaning." Likewise, the bizarre American basketball player Dennis Rodman once declared, "Fifty percent of life in the NBA is sex, and fifty percent is money." Now and then they also play basketball.

In the animal world, biologists have long known that males compete primarily for the attention of females and that females, almost against their own better judgment, often fall for the winners, the ones who control the resources. On the California coast, for instance, northern elephant seals wage spectacular and sometimes bloody battles to establish control of a few prime beaches. The winner stays on his beach for an average of ninety-one days without going back to the water to eat, ruthlessly defending his territory even as he loses a thousand pounds of stored fat, roughly 41 percent of his body weight. He does it because the beach is where the females wallow in all their blubbery, lubricious splendor, and a male who successfully defends his territory typically gets to have sex with thirty of them in a season. Meanwhile, the vast majority of lesser males never copulate at all. In one study of 115 males, the top 5 males sired 85 percent of all offspring. Likewise, most studies of baboons have found that high-ranking males have sex significantly more often, or with more females, than do subordinate males. Social dominance is so sexually compelling that some female ducks engage in the ultimate kinky behavior, bypassing their own middle-rank males to mate with high-status males of another species. (Well, the lives of the rich should probably teach one never to use the words "ultimate" and "kinky" in the same sentence: The time-traveling actress Shirley MacLaine once bypassed all living males of *all species* to have sex with the dominant but very dead Emperor Charlemagne.)

Sporting in the Cinnabar Crevice

At about the same time Sonia and Amir were beginning their marriage, an anthropologist named Laura Betzig, now at the University of Michigan, set out to test whether high status translates into reproductive success in the human world, too. She didn't expect to find much. Betzig was a product of a Midwestern culture in which monogamy was deeply entrenched. Her father, a successful machine tool manufacturer, had been married to her mother for thirty-five years. Then he read his daughter's thesis and got divorced.

What Betzig found in extravagant detail was that, at least until modern times, men have almost always used wealth and power to get as many women and as many offspring as they could possibly afford. Julius Caesar's soldiers celebrated his womanizing in verse ("Home we bring our bald whoremonger; / Romans, lock your wives away!") and legislation was drawn up legitimizing his union, for the purpose of procreation, with any woman of his choice. He chose, among others, the wives of Pompey and Crassus, his partners in the first Triumvirate, the cuckolding of friends and rivals being a particularly piquant way to reinforce one's rank in the hierarchy. Augustus, celebrated by classical scholars for devotion to his third wife Livia, liked to deflower young girls procured for him by Livia herself. Other Roman noblemen kept hundreds or even thousands of slaves, often women chosen for their ability to bear children. Chinese emperors employed courtiers to schedule their sexual encounters for days in the month when their consorts were most likely to conceive. Indian rulers kept seraglios and the "compassionate" ones mustered the sheer moral fiber to do the right thing by at least two women per night. Even in the more restrained circumstances of Christian Europe, Charlemagne had five successive wives and four concubines. Half of modern Europe can thus claim him as an ancestor.

Sleeping with a multitude of women was not merely a privilege of rank but also at times a sacred obligation. Like ancestor worship, though rather more agreeably, it was a way for a member of the Chinese aristocracy to enhance his *tê*, or family magic. In any case, this seems to be what aristocrats told their wives. The *Tso-chuan* chronicle, dating back about 2,500 years, includes an aphorism that resonates in the houses of the rich to this day: "The *té* of a girl is without limits, the resentment of a married woman is without end." The Chinese aristocracy regarded young women as wellsprings of *yin*, and believed that the best way for a man to supplement his precious *yang* was not merely to sleep with as many of them as possible, but also "to prolong the coitus as much as possible without reaching orgasm; for the longer the member stays inside, the more *yin* essence the man will absorb, thereby augmenting and strengthening his vital force." To relieve the tedium of *yin*-gathering, guidebooks for the upper classes recommended such positions as "Phoenix Sporting in the Cinnabar Crevice," "Wailing Monkey Embracing a Tree," "Reversed Flying Ducks," and "Hounds of the Ninth Day of Autumn." Should any of this actually bring a man too abruptly to the brink of ululation (or, as an ancient Chinese sexologist might have put it, "Baying at the Moon with Toes Curled"), another guidebook advised that "he should quickly press the *P'ing-i Point* with index and middle finger of his left hand, then let out his breath, at the same time gnashing his teeth a thousand times." Such were the sacrifices selfless rich men made for the good of their families.

The Swarm of a Rich Man's House

Betzig argued that behavior formerly regarded as the depraved exception was the rule among the wealthy and powerful, beginning in our hunter-gatherer past when men who brought home the most meat won the most mates and continuing on at least through the heyday of the British aristocracy (to say nothing of Woody Allen). This was one other reason the houses of the rich needed to be so big: To store spare women. Among the most extreme cases of what might be termed "the honeypot hypothesis," Betzig cited Aztec Mexico, where Montezuma kept four thousand concubines; Inca Peru, where emperors kept multiple "houses of virgins" with fifteen hundred women in each; India, twenty-five hundred years ago, where an emperor is said to have kept sixteen thousand women; and, her record case, Imperial China, where the harems numbered up to forty thousand women. (On the Great Basketball Court in the Sky, Wilt Chamberlain weeps.)

On a somewhat less ambitious scale, ordinary rich people also customarily retained household servants, and up until World War II, a retinue of thirty or forty was not unusual. In the big houses of the British countryside, Betzig wrote, these servants "consistently shared three traits: they tended to be young; they tended to be female; and they were almost always unmarried." At least in the minds of many upper-class males, "Housemaid Heights," as the female living quarters at Blenheim were known, constituted a harem rightfully theirs for the raiding. Samuel Johnson, the lexicographer, argued that a wife "ought not greatly to resent" it if, "from mere wantonness of appetite," her husband "steals privately to her chambermaid." James Boswell, the diarist and Johnson's biographer, also happily rationalized the philandering impulse: "If I am rich, I can take a number of girls; I get them with child; propagation is thus increased. I give them dowries, and I marry them off to good peasants who are happy to have them. Thus they become wives at the same time as would have been the case if they had remained virgins, and I, on my side, have had the benefit of enjoying a great variety of women."

Many masters, or their indignant wives, simply fired their "disgraced" servant girls. An 1883 investigation by the Registrar General in Scotland found that domestic servants accounted for almost half the births out of wedlock and a census in midcentury concluded, "A county with more servants had more bastards." A girl fired for being pregnant typically received no reference letter and thus had little prospect of employment other than in what Boswell blithely termed "Paphian bliss"—that is, prostitution.

Coercion was undoubtedly a factor in many, if not most, master-servant liaisons. Betzig cited the 1743 self-help pamphlet *Present for a Servant-Maid,* which offered this sober advice about fending off the master: "Being so much under his Command, and obliged to attend him at any Hour, and at any Place he is pleased to call you, will lay you under Difficulties to avoid his Importunities." The book advised the maid to persevere and scold her master for sinfulness. (Jonathan Swift, a clergyman, offered more worldly advice: "Make him pay . . . Take care to get as much out of him as you can; and never allow him the smallest liberty, not the squeezing of your hand, unless he puts a guinea into it.") But negotiations weren't always an option. In her eighteenth-century memoir, *An Apology,* Con Phillips described how her career as a courtesan began. Philip Stanhope, heir to the third earl of Chesterfield, invited her to his rooms to watch some fireworks; then he plied the thirteen-year-old virgin with liquor, tied her to a chair, and raped her. (While Con Phillips was becoming a courtesan, Stanhope went on to become the moralist Lord Chesterfield, whose sayings include: "Know the true value of time; snatch, seize, and enjoy every moment of it . . . never put off till tomorrow what you can do today." And also: "Sex: the pleasure is momentary, the position ridiculous, and the expense damnable.")

Subtler forms of coercion may also have worked. In her memoirs, Consuelo Vanderbilt described an intriguing incident in which her husband, the ninth duke of Marlborough, accused a maid of stealing a small china box. "The housemaid supposedly responsible was sent for [by the head housekeeper] and, crying, she said she wished to leave as she had never before been accused of stealing. When at last I managed to soothe them both I went to see Marlborough, who laughingly informed me that he had himself hidden the box to see if they would notice that it was no longer there." Vanderbilt attributed this incident to her husband's "fastidiousness," though other motives for putting a maid at a disadvantage must surely have come to mind. Only a few years earlier, the duke's own uncle (and Winston's father) Sir Randolph Churchill had died of syphilis ostensibly acquired from a Blenheim maid.

Though it may be unpalatable to feminists, the other possibility is that women often succumbed in such situations because they wanted to. That is, they accepted a sexual partner because they were powerfully attracted to him as a dominant male, because they saw what looked, at least at the moment, like their own best economic and social advantage, and even perhaps because of love.

Crossing the Polygyny Threshold

Long-billed marsh wrens are mousy brownish birds found in dense reedbeds and cattail stands through much of North America. The male arrives first in the spring, establishes a territory, and defends it with the usual masculine bluster and belligerence. He builds a group of incomplete nests amid the reedstems to attract a female. Then he advertises his great achievement with a sort of song-and-dance. If a female accepts him as a mate, she picks one of the nests and feathers it with cattail down. Then she settles in to raise a family. At this point, the male's attention begins to roam. He wanders off to build a new cluster of nests, as a courtship center for attracting another female. Biologists studying the species in the 1960s noticed that males with better territories tended to attract more than one female. In fact, some males had as many as five females, even when there were plenty of frustrated bachelors whiling away their lonely hours and singing *a capella* on every street corner in the neighborhood.

The biologists described a phenomenon they called "the polygyny threshold," *polygyny* meaning "many females." It's the point when females recognize they are better off sharing a male who already has a mate than looking for a full-time male of their own. A male with a number of females to attend will necessarily be a road warrior and spend only quality time, if that, at each nest. But the female may knowingly choose to become second or third bird in a harem, according to one recent research paper, if the male possesses sufficient resources to "compensate the female for any cost of polygyny." That is, he needs to be rich. ("How much is she evaluating the male, and how much is she evaluating the real estate?" asks Gordon Orians, who has studied female choice in these circumstances. "In redwing blackbirds, we think the real estate is more important than the male.") The polygyny threshold is now known to exist not just in marsh wrens but in starlings, blackbirds, damselflies, bark beetles, and many other species.

The idea that a polygyny threshold also exists for humans is, on the other hand, highly arguable. Along with randy womanizers, the ranks of the rich have often been filled with dour sorts like John D. Rockefeller, Sr., or H. Ross Perot, who seem to accumulate status and resources for their own sake, not primarily to catch a woman's eye, much less the eyes of many women. Likewise, television talk show host Oprah Winfrey does not appear to have piled up her millions for the chief purpose of attracting boy toys. You could easily argue that Armand Hammer expressed a more characteristic attitude among rich people when he said, "Money is my first, last, and only love."

Modern humans have a strong cultural and legal bias in favor of monogamy; personal experience can also make it seem like a healthier way of life. Rockefeller, for instance, became a devoutly faithful husband in part to compensate for his father's philandering (much as his own son would become a philanthropist to compensate for Rockefeller's reputation as a predatory businessman). William Avery Rockefeller, a wealthy patent medicine salesman, had kept wife and mistress in the house where Rockefeller was born, sired illegitimate daughters on either side of John D., was once indicted for raping a housekeeper, and eventually abandoned his family. John D. reacted by becoming "a sweet respectful Victorian husband" to his only wife Cettie and by actively involving himself in rearing their children.

Yet our propensities have a nasty habit of reasserting themselves. Despite his best intentions to make money his only love, for instance, Armand Hammer was supporting at least two current mistresses when he died at the age of 92, plus a daughter who'd been conceived by another mistress at roughly the time he was marrying his third wife. When Larry Hillblom, founder of DHL Worldwide Express, died in a 1995 plane crash, it came out that he had a special fondness for teenage virgins in Asia. Four of his children by separate unmarried mothers successfully sued for the bulk of his $650 million estate. And (here we leave behind all precedents in nature and veer off into the sweet human surreal) Hillblom's disapproving mother, who'd been left out of his will, sold a sample of her blood to the plaintiffs for $1 million to help them prove that Hillblom was the father. According to another recent lawsuit, the Sultan of Brunei and his brother Jefri (whose 181-foot yacht, "Tits," is equipped with speedboats named "Nipple One" and "Nipple Two") used some of their billions to import a rotating harem of beautiful women. For a fee of $3,000 apiece per day, the women, including a Penthouse Pet and a Playboy Playmate, were encouraged to perform karaoke, and other indecent acts.

The Love Infrastructure

The footpaths to infidelity are part of the infrastructure of a rich man's life. Friends and faithful servants often work to ensure that the boss has plenty of female company and also to spare him the pain of rejection. Queen Victoria's son, England's leading voluptuary in the decades before he became King Edward VII, would simply "eye the audience from his box in the theatre and send an equerry to invite the most attractive woman to join him. It was an offer that was rarely refused." Rank gave him a sort of *droit de seigneur* to take away females from lesser males. (No doubt other male theatergoers were secretly delighted when "one night in Paris the subject of his interest proved to be Prince Yussopov, later to be Rasputin's assassin, who was in the habit of dressing in women's clothes." But at least the use of his equerry spared the prince from making this blunder face to face.)

Procurement is still frequently part of the unwritten job description of personal assistants to the rich. "You spend a lot of time together," a personal assistant remarked, over dinner one night in Los Angeles. "You see them not just in the office, but in the middle of the night. So it is personal."

"And it is *highly* confidential," said her friend, who added that her former boss was too rich (and too ugly) to go out cruising night spots for romance. "He once offered me a $1,000-a-month raise to get him a date with a woman at the law firm down the hall." The relationship lasted three months. "He also offered me $10,000 to get him a date with Sigourney Weaver. We offered her a part in a commercial that didn't exist." Weaver's staff wisely screened that call. Van Halen front man David Lee Roth used to give each member of his crew five backstage passes to initial and distribute to likely women. He then checked the passes of the women who ended up in his bed and paid $100 to whichever crew member had initialed the back.

Hiring prostitutes is at times also part of the job. "They don't call them that," said another West Coast assistant. "I hire 'dancers.' So whatever happens after that, it's between them." Bimbo control figures largely in the life of a personal assistant to a National Basketball Association star: "She arranges abortions, sets up dates, goes to Bloomingdale's and buys these little gifts," says the hiring agent who got her the job. "She keeps track of what he wants told to the girls. He may have told one woman he's out of town, when he's going out with another woman." The assistant describes herself as "the only woman in his life that he doesn't try to sleep with."

Mate-Copying

Apart from active procurement, another factor making polygyny easier for the rich is imitation. "The more women you spend your time with and the more public it is," *Playboy* founder Hugh Hefner told me, a little smugly and a little puzzled, "the more other women want to be somehow involved. I think it has a tremendous sexual power, and a sexual appeal." It's possible that they are all responding independently to the same animal magnetism—but not even Hefner would make that argument. He volunteered that he had to put on the trappings of the playboy, the smoking jacket and the pipe, before he really became the persona his magazine was already marketing. Women on the arm were simply one more trapping.

Arguably the most effective trapping: Arm candy gets the attention of other females the way a decoy attracts ducks. In 1990, a biologist named Lee Dugatkin made a series of experiments with two female guppies and two males. He placed one of the females, dubbed "the model," with one of the males, and he allowed the other female to observe the happy couple together, and the lonely guy off by himself. When he removed the model, the observing female almost always made a beeline for the model's happy boyfriend, and not for the lonely guy. When Dugatkin reversed the experiment and put the model with the former lonely guy, the observer now almost invariably chose the erstwhile loser. Her choice apparently had little to do with the quality of the particular male. The observer was simply "mate-copying" the choice of the other female.

Roughly the same thing apparently happens in wild populations of sage grouse: Males in this North American species gather in a lek or display area, a sort of sage grouse singles bar. One male in a lek of perhaps twenty males typically gets 80 percent of the matings. In slow years, when a female doesn't have much competition, she seems to choose a male based on some individual trait she considers desirable. But in busy seasons, things snowball in an odd way: The more other females see a particular male mating, the more they line up to mate with him—often five or ten of them in a day. They don't seem to give a damn whether he has desirable traits or not. Mate-copying by itself makes him a hot item.

In other words, females may choose a male by either of two strategies: They may go through the process of rigorous scrutiny we call courtship, or they may simply imitate the mate choice of other females as a shortcut for finding desirable males. The idea that mate copying also exists in humans is unsurprising to anyone who has survived high school dating. After his experiments with guppies, Lee Dugatkin and colleague Perri Druen set out to analyze the logic of what they call date-copying in humans. They asked participants in a survey to rate the attractiveness of an individual based only on physical appearance (the classic Bo Derek "on a scale of one-to-ten") and the knowledge of how many members of the opposite sex, out of a group of five, actually wanted to date the subject. It turned out that both males and females cared about popularity, which suggests that date-copying is an equal-opportunity behavior. But "the most interesting finding in our date-copying experiment was not how important popularity was, but what traits were attributed to popular individuals. . . . What we discovered is that while both males and females attributed social skills, a sense of humor, and wealth to popular individuals, a critical difference between the sexes emerged: Our female subjects clearly stated that they assumed if others were interested in a particular male, that male was wealthy. Why they made that leap we can't say, but that they made it is unambiguous."

Attitudinal surveys are a tricky sort of science, ripe with the potential for cultural stereotyping of all sorts. But with that caveat in mind, they are also kind of fun. Other attitudinal surveys suggest not only that men gain status from their arm candy but also that the highest status accrues to an ugly man with a beautiful partner, like Aristotle Onassis with Jacqueline Kennedy. The assumption people seem to make, when you boil it down, is that an ugly man with a stunner on his arm must be extremely rich. The awful implication of this is that a beautiful wife, dressed in haute couture and important jewelry, may unwittingly serve as a sexual advertisement for her husband.

The War on Copycats

In the natural world, the primary female is frequently distressed by the whole notion of polygyny. Her road warrior mate provides the same amount of parental care as would a monogamous male, but this care now gets divided up among multiple broods. To keep him working for her offspring alone, female starlings in one study went on the warpath against any "prospecting female." A bit like Ivana Trump getting into a public shoving match with Marla Maples when Donald Trump made the blunder of bringing them both out to Aspen at the same time. With starlings as with starlets, the primary female may also copulate more frequently with the wandering male, in an attempt to render him *hors de combat*. Male starlings counter by withholding sex. The smarter ones also set up their love nests at a prudent distance from home. The primary female probably isn't fooled by her road warrior's deceptions. She merely makes the best of a bad situation: at the end of the day, alone in her nest with her yawping brood, she still at least has the real estate.

Getty & Son Enterprises

The polygyny threshold has played a formative role in the history of many wealthy families, though they have probably never heard of the term. It is part of their unwritten heritage. In her novel *The Edwardians*, Vita Sackville-West depicted the ghosts of the mistresses kept over hundreds of years by the lords in one country house as a "charming cohort" who "peopled the corridors and insinuated their suggestions into ears well attuned to listen. If the fifth duke had made a scandal in the reign of Queen Anne, why shouldn't his Grace make one now, if he was so minded?"

Nor was this purely the stuff of country house fiction. At the age of thirty-four, for example, J. Paul Getty married his second wife, Allene Ashby, while still married to his first wife. At the same time, he was having an affair with Allene's sister Belene. Getty also married his fourth wife, Ann Rork, in bigamous circumstances. When Rork, who was already pregnant, joined him in Europe shortly thereafter, Getty celebrated her arrival by going out on a date with another woman. According to Getty biographer Robert Lenzner, Rork used to unpack Getty's luggage when he returned from trips and find "his contraceptives and other sexual paraphernalia. According to Ann, he even showed her a one-page legal agreement . . . which he asked women to sign before he bedded them. It waived any claim against him if they became pregnant." Perhaps this was to reassure her that, apart from the sons he had already fathered by wives one and three, she need not worry about his attention being diverted to other broods. Though by then, Rork probably already knew that Getty was constitutionally incapable of paying attention to any of his offspring. Lenzner continues: "He liked to try to pick up women on street corners or other public places, to see if they would go to bed with him without knowing his identity. These anonymous assignations were to prove to himself that he could be loved for reasons other than his name and his wealth. Sometimes, when he and Ann went to the movies, he would arrange for his old flame, Belene Ashby, to meet them. He liked having more than one woman admirer around." Getty expressed a rich man's view of marriage this way: "A lasting relationship with a woman is only possible if you are a business failure."

Gordon Getty, the second child of the marriage to Ann Rork, gave every indication of taking the opposite path: "In total contrast to his father, he was naturally uxorious," another Getty biographer declared in 1989, "and his marriage proved the rarity among the Gettys—a stable and happy relationship." This may well have been true, though perhaps not in the way the author intended. Gordon and his wife, Ann, the daughter of a fundamentalist Baptist walnut farmer, have now been married for thirty-eight years and have four grown sons. They dominate the social scene in San Francisco, where Gordon has made a reputation as a philanthropist, a critically disclaimed opera composer, and a shrewd though reluctant businessman: When his father was alive, Gordon fought him for a fair share of the empire. Later, with impeccable timing, he arranged the sale of the family oil company for $10 billion. His own and his relatives' share came to $4 billion, instantly increasing their wealth by more than J. Paul Getty had earned in his entire lifetime.

Gordon Getty was unlike his father in some ways, and in others much more like him than he let on. His relationships with women were lasting, but he also managed to maintain more than one at the same time. In 1999, it came out that he had in effect kept two wives and raised children with both of them. Getty had set up his love nest at a prudent distance from home, in Los Angeles. He flew back and forth in the family 727, the Getty "Jetty." At an inflight cost of $10,000 an hour, this made his polygyny rather more expensive than that of the average marsh wren, but expense was hardly an issue. His involvement with the construction of the Getty Center, his father's museum, took him to Los Angeles as often as three times a week, and there, at a party for the center, he met Cynthia Beck, from one of the few genuinely old money families in Los Angeles.

The resulting romance was said to be common knowledge within the discreet circles of the rich. It became public knowledge, too, when Getty's three daughters by Beck filed legal papers to change their last name to Getty. "I love my father, and I want the world to know I'm his daughter," said the eldest of them, a fourteen-year-old. This must have been distressing news not just to Getty's wife but to their four sons: Seven does not go nearly as well as four into $1.5 billion, the estimated worth of Getty's estate at that point. Their father declared that "the Getty family has been fully supportive throughout this situation." Ann Getty made no comment. "Wait six months," a divorce lawyer told one gossip columnist. "There will be a discreet announcement that Gordon Getty now has a Los Angeles address." But Ann and Gordon Getty remain married three years later. Among other social events at which they have appeared, the pair graciously hosted a friend's San Francisco wedding in mid-2001. "When the wedding party came down the aisle," a local newspaper reported, practically quivering with pleasure, "Gordon Getty took the hand of his wife Ann in his own."

One does not have to seek far among the rich to find similar cases. Sports figures including Roscoe Tanner, Pete Rose, Steve Garvey, and Jim Palmer have all paid support for children born out of wedlock. King Albert II of Belgium has fathered an illegitimate daughter by a French baroness. The octogenarian head of one of Europe's great banking families recently conferred his family name on the daughter he fathered in a long-term extramarital affair with an antiques dealer. He's still sometimes seen at parties with the antiques dealer and still also married to his wife. In such cases, it's seldom in the husband's interest to divorce. "In certain circles, it's classy to still have the original wife," one art world regular told me, and another immediately added, "They tend to be richer."

Rich wives may also have no better option than to arrive at an understanding and acquiesce in a husband's polygyny. They could certainly divorce and come away with one or two fine homes and a potful of money. But the homes and money are already theirs, and an older woman in particular is unlikely on her own to retain the status in her circle that she enjoys as a rich man's wife. The rich octogenarian's wife, for instance, appears to find his infidelity vexing; she once smashed her car into the car of one of her husband's former lovers. But her marriage makes her a baroness, an active member of some of the world's most important art museum boards, and also one of the most respected women in Europe. Likewise, Ann is a Getty, not a walnut farmer's daughter. Moreover, marriage keeps their connection to their children and the lineage intact.

None of this sounds wonderful for the consorts of rich men. In the long chronicle of their humiliations, for instance, King Edward VII's wife Alexandra had to pretend not to notice that one of his favorite horses was called Ecila, the name of his mistress, Alice Keppel, cunningly spelled backward. (On the other hand, Alexandra liked to refer to another mistress, Jane Chamberlain, as "Chamberpots.") More recently, a middle-aged Hollywood producer from a venerable family was living with a beautiful, but unbright, young woman. He hired a local academic to make her presentable, and one day tutor and student were sitting around discussing what to get their boyfriends for Valentine's Day. The tutor's plans were conventional, a book, a nice shirt, some theater tickets. Then the student announced her plan: "He already has everything. So I'm going to get him Stephanie."

La Donna è Mobile

Yet a closer look at the polygyny threshold is enlightening about the female point of view. In the past few years, DNA analysis of the offspring from such partnerships has produced surprising results. They demonstrate that harem females, and particularly the junior females, often slip away for extramarital flings of their own. In one study of starling offspring, 40 percent of all females had nestlings fathered by somebody other than the ostensible mate. Almost 16 percent of all nestlings were EPY, or "extra-pair young."

Biologists now believe that infidelity is entirely natural for females. If evolution inclined Sarah Ferguson to marry Prince Charming, for instance, evolution, and the prospect of lifelong monogamy with a Windsor, also gave her powerful incentives to seek toe-sucking on the side. Having secured a good provider, a woman may see fringe benefits, in the attention of lovers. Likewise, swallows, which are ostensibly monogamous, get assorted gifts, attention, and about a third of their offspring from males other than their mates.

Given the enormous cost invested by the female in every egg, and given also that she hopes to induce her mate to invest his energy in caring for their young, why would she risk it all by messing around with passing strangers? The proximate factors are familiar: It feels good and, since her husband's a road warrior anyway, she can probably get away with it. But biologists suggest that two "ultimate factors" are what really make infidelity a useful Darwinian strategy for females: Fooling around provides "fertility insurance" where the primary male is sterile or impaired, a common enough circumstance in the case of beautiful young women married to older men; and it gives the female a chance to attract "good genes"— perhaps better genes than her husband's. In his book *Promiscuity*, the British behavioral ecologist Tim Birkhead gives the example of a swallow who gets delayed on her migration home from Africa. With all the other males already taken and her biological clock ticking, she decides to make do with Mr. Average Swallow. They build a nest together, allowing her to breed in the first place. But "in the middle of her fertile period she gives her partner the slip, finds one of the really attractive males, and copulates with him." The result is that she rears legitimate and illegitimate young in the same brood, the latter probably more attractive than the former to future mates—and thus more likely to give her grandchildren.

Again, one does not have to look far to find the equivalent among wealthy humans. Harold Macmillan, from an upper-middle-class British publishing family, was posted to Canada after World War I as aide-de-camp to the Duke of Devonshire, then serving as Governor-General of Canada. There he fell in love with the boss's daughter, Dorothy Cavendish. She came from a line of rich aristocrats who "were traditionally drawn to success in whatever field," according to biographer Alistair Horne. "Cavendish women also had a reputation for being highly sexed." Macmillan on the other hand was "an enthusiast who does not enthuse," "dowdily prosaic," with "a sense of devotion bordering on the dutiful." Not too surprisingly, Dorothy was soon swept away by her husband's friend Robert Boothby, who "dashed about in an open, two-seater Bentley and was socially at ease wherever he went." Macmillan, who went on to become British prime minister, remained married to Dorothy, but Boothby apparently fathered their daughter Sarah.

Unfortunately, no reliable numbers exist for the average percentage of extra-pair young in human populations. Estimates based on blood tests have shown that false or misassigned paternity ranges from 2 percent among the !Kung of Botswana to as much as 30 percent in the "Liverpool flats" of northern England. Much as with Mr. Average Swallow's wife, participants in the British study reported that extra-pair copulations were significantly more likely to occur just before ovulation. That is, getting good genes seemed to be part of the strategy, at least at some level. Unfortunately, that study reported on the opposite end of the economic spectrum from the rich, who do not by and large line up to volunteer evidence about whether they or the offspring happen to be bastards.

Romans, Lock Your Wives Away

Rich men have an ample history of attempting to prevent their wives from stepping out on them with other men. In the natural world, this sort of behavior is called mate-guarding, though a woman I know in New York refers to it somewhat more plainly as "cock-blocking." Animals do it by a variety of highly creative means. People who see dragonflies flying in tandem, for instance, often mistakenly assume that they are witnessing a winged orgy. But it's usually just desperate male clinging. After inseminating a female, the male hangs onto her as long as he can, following her down to the water and hovering as she lays her eggs, to prevent other males from sneaking in and displacing him as the father. The dungbeetle actually rides like a pasha on the female's back during egg-laying, and the bearded weevil straddles the ovipositing female with his long legs to imprison her. If a rival male approaches, he shoves his long shovel-like snout underneath the newcomer and flips him unceremoniously onto his back. (The human equivalent is to reduce a rival to a laughingstock, in the manner of the Earl of Sandwich. Admonished by a moralizing politician that he would die either of the pox or on the gallows, Sandwich replied, "That depends, sir, on whether I embrace your mistress or your principles.") Certain insects, snakes, and rodents actually manufacture a quick-setting copulatory plug to block the genitalia and prevent the female from indulging in any future mating. Think of a chastity belt made of Krazy Glue.

Wealthy human males have also sometimes resorted to physical impairment as a form of mate-guarding. In China, footbinding wasn't simply a way to demonstrate that a wealthy woman was above the necessity of doing physical labor; it was also a way to keep her immobilized. Harems guarded by eunuchs were of course also intended to discourage other males. In some cultures, clitoridectomy served to prevent a woman from achieving sexual pleasure, thus making her husband more certain that he was the father of her children. Other countries developed manufactured chastity belts, and, odd as it may seem, people still apparently buy them today. They cost about $500, and in December 2000, the British magazine *Tatler* declared them "the season's key accessory." Tongue, one hopes, locked firmly in cheek.

Some evolutionary psychologists theorize that mate-guarding tends to increase with wealth, because rich men have more to lose by a partner's infidelity: Martin Frankel, the insurance swindler, used his money in the late 1990s to establish a harem in Greenwich, Connecticut. One mistress complained that Frankel hired minders to discourage her from infidelity, and a guard once threatened her with an electric cattle prod. The households of other wealthy people are generally more subtle than that. Armand Hammer, for instance, merely placed a tracking device on a lover's car and wiretapped her phone. Men of more magnanimous disposition make a show of cosseting their wives and lovers with household staff, who function as *de facto* mate guards. "Indeed, with a page in the house, a coachman or a postilion to take me for drives and a groom to accompany my rides," Consuelo Vanderbilt remarked in her memoirs, "my freedom was quite successfully restricted." Her husband's motive for guarding her in this fashion becomes evident (to the reader, though not apparently to Vanderbilt herself) later on the same page, when she notes one of the duke's rare, backhanded compliments on her beauty: "If I die," he told her, "I see you will not remain a widow long."

The vestigial remnant of mate-guarding in modern society, where a wife tends to resent being kept prisoner in her own home, is for the man simply to check in randomly on the woman's cell phone. It's also possible to recall the phone log and see whom she's been calling. When she was married to Ronald Perelman, Patricia Duff sometimes borrowed cell phones from friends to prevent her husband from tracking her every move.

Sonia and Amir

Full-blown mate-guarding also persists, notoriously, in many Muslim countries, like the one to which Amir took his new wife Sonia. (For her own safety, she will not allow the country to be named, though she no longer lives there.) Like all women, Sonia was obliged to cover herself with veil and gown. But her husband's wealth gave her the means to buy Paris fashions, and she did so compulsively, at the rate of twenty-five suits or dresses a season, costing about $1 million a year. "I could have any dress I wished, any piece of jewelry I wished. But I couldn't wear them because I wasn't allowed to go out in public." The exception, she said, was when she was pregnant "and no threat to him."

The rules also relaxed, but not for long, when they traveled in Europe. One evening at a fashionable restaurant in Monaco, a man "came over to say hello, telling me how pretty I looked, chatting about this and that without paying attention to my husband. I mean, how do you control an Italian man in front of an elegant young woman? Amir couldn't stand it: we left the restaurant without eating." They returned to their yacht, where her husband went down to her dressing room, took a pair of scissors, and cut up every fashionable dress that came into his sight.

This sort of thing may well discourage infidelity among wealthy women. The evidence is scant, but in one study, wealthier women reported having had fewer sexual partners, the opposite of the pattern in wealthy men. Wealthier wives also said they were less likely to have been unfaithful. This double standard may well get reinforced because, regardless of wealth, women lose sexual status as they grow older, whereas wealthy men seem to retain their status even, like Armand Hammer, unto decrepitude. Another interpretation is that wives of wealthy men are less likely to have affairs—or at least less likely to admit to them—because "they have more to lose if they get caught and are deserted by their husbands." When Sonia eventually separated from Amir and began what she maintained was a chaste romance with an American, her husband simply reminded her that in his country, where he held their five children, the penalty for an adulterous wife is death.

Sneaker Males

Rich men distracted by business affairs of one sort or another often entrust their wives to suitable male escorts, or "walkers," the modern equivalent of eunuchs in the harem. It's a sign of how important mate-guarding remains that "suitable" in this context almost always means homosexual. But life for the rich, and other animals, may not be as reassuringly simple as this suggests. The natural world is full of stray males, called "satellite males" or "sneaker males" by biologists, and together with willing females, they make it their business to defeat the expectations of mate-guarding males.

In the context of the society walker, for instance, consider the deceptive sexual strategy of certain transvestite stickleback fish. When one of the more "manly" sorts of sticklebacks wants to attract potential mates, he scoops a shallow nest out of the sandy bottom and invites females to come in and lay their eggs, on which the male then spills his seed. Some other males don't bother digging nests. They simply adopt the drab coloration and characteristic body movements of females and thus, like a rich woman's walker, they get invited into the nest—whereupon they beat their host to the act of fertilization.

Males who do not fit the alpha stereotype pursue a similar strategy in species as different as ruffs and white rhinos. And among wealthy humans? Most older women seem to like their walkers to be genuinely homosexual, because it is simply less complicated. All such women usually want, as one of them put it, "is a walking-stick." Even a younger woman like Ivana Trump was afraid, at the time of her divorce, to be seen in the company of any man not known to be gay, lest her husband charge her with adultery. Yet hints of sneaker-male behavior occasionally flash to the surface. Jimmy Donahue, an heir to the Woolworth fortune, was openly homosexual long before it was socially acceptable. He once gave a luncheon for the Marquess of Milford Haven and asked afterward if his guest would care to see his mother's "famous bronze collection." Then he opened a door to reveal "half-a-dozen black men, oiled and standing naked in provocative poses." So no one in society initially believed his relationship with Wallis Simpson was physical. But according to author Christopher Wilson, Donahue and Simpson carried on a passionate sexual affair, to the chagrin of her dysfunctional husband, the Duke of Windsor. Evan Frankel, the flamboyant Broadway producer, was known among women for groping and fondling—but also managed to be suspected by men of homosexuality. He sometimes played the part of the walker, and one of the rich wives he escorted recently remarked: "Evan loved the idea that people were going around saying he was gay while he was secretly screwing all these young girls—and everybody's wife." Princess Stephanie of Monaco has also demonstrated a consistent preference for sneaker males, though the sexual stereotyping is the reverse of what seems to happen among sticklebacks. She has always had access to the wealthiest and most aristocratic men on Earth, but her first two children were conceived in a relationship with her bodyguard.

Sexual Advertising

Like Princess Stephanie, young females in the natural world sometimes make the mistake of going public when they have sex with a sneaker male. This tips off the alpha, who typically comes crashing through the brush to drive off his rival. (In Stephanie's case, her unsuitable liaisons appear to have been one of the factors that provoked her father to cut her inheritance to just 1 percent of his estate.) In his book *Chimpanzee Politics*, Frans de Waal describes a female named Oor, who used to scream with unusual volume at the point of climax with any male, until she figured out that this wasn't always such a bright idea. "By the time she was almost adult she still screamed at the end of mating sessions with the alpha male," but hardly ever during her clandestine "dates" with beta males: "During a 'date' she adopted the facial expressions which go with screaming (bared teeth, open mouth) and uttered a kind of noiseless scream (blowing from the back of the throat)." Oor had apparently figured out how to use her vocalizations to her own strategic advantage.

Like Oor, human females also appear to vocalize differently with different males, and they may willingly participate in sexual advertising when affiliation with a dominant male also confers status on them. Marla Maples served Donald Trump a little too enthusiastically in this line, when she made the front page of *The New York Post* by confiding to an acting school classmate that Trump had given her "the best sex I've ever had." But going public about sex with an alpha, and keeping quiet about sex with a beta, is normal behavior even outside the worlds of Donald Trump and Oor the Chimp. Queen Marie Thérèse, for example, used to announce that her husband King Louis XIV had slept with her, rather than with one of his mistresses, by clapping her hands when her court came to attend her in the morning. Pamela Harriman apparently enticed Elie de Rothschild at least in part because he knew he would be following in the footsteps, as it were, of a Churchill, a Harriman, a Whitney, and an Agnelli. Likewise, Lillie Langtry paraded around London on sidesaddle in a skintight dress with the Prince of Wales but did not let anyone, least of all the prince, know that she was also having an affair with a childhood chum named Arthur Jones.

Corridor Creepers

The researchers who found such high percentages of extra-pair young among starlings didn't identify the real fathers. Straying females may well have gone slumming with some of the hapless bachelors in the neighborhood. Or they may have indulged in their little trysts with ostensibly monogamous males, that is, males who maintained only one nest site. But the researchers believed that straying females were most likely to have sought out other polygynous males, a case of the rich getting richer. Studies in other species also suggest that females are more likely to go up-market for their infidelities. One long-term observer of black-capped chickadees found, for instance, that in every case of extra-pair copulation she witnessed, the female sought out a male who was socially dominant over her own partner. In human terms, females are less likely to stray with some harried middle manager than with a J. Paul Getty or a Larry Ellison.

The British aristocracy appear at times to have built an entire mating system on these lines, with women evidently in control. This was the cause to which the country house weekend was largely devoted, a point Arnold Bennett seems to have missed when he complained, after one such weekend at Blenheim Palace, "The bulk of the twenty guests seemed to me to have no interest whatever except sexual." One must add in their defense that this interest seems at least to have been genuine and enthusiastic. These weekends were ostensibly organized for slaughtering large quantities of game birds, specially fattened for the occasion, or for chasing foxes hell-for-leather cross country. But another kind of hunt took place in sitting rooms and around dinner tables (and sometimes also in the field). "Among the most sophisticated of these women, often bored partners in arranged marriages, the affairs which were joyously celebrated during weekends were sometimes launched in wife-to-wife conversations," William Manchester wrote of the fast crowd gathered around the Prince of Wales in the late nineteenth century. " 'Tell Charles I have designs on him,' one would tell Charles's lady, who would acknowledge the proposal with a nod and an amused smile; she herself already had a lover or had designs on someone else's husband. But you had to be very secure to take that approach. . . . More often an understanding would have been reached in advance between the primary partners."

On one such weekend, a hundred guests slept at Blenheim, and exactly where they slept was a subject to which the lady of the house applied all her considerable diplomatic talents. "This question of the disposition of bedrooms always gave the duchess and her fellow-hostesses cause for anxious thought," Vita Sackville-West wrote in *The Edwardians*. "It was so necessary to be tactful, and at the same time discreet. The professional Lothario would be furious if he found himself in a room surrounded by ladies who were all accompanied by their husbands. . . . It was part of a good hostess' duty to see to such things; they must be made easy, though not too obvious." The name of the guest in each bedroom was elegantly written on a card placed in a small brass frame on the door. After the party had retired to their rooms for the night, gentlemen known as "corridor creepers," the aristocratic euphemism for "sneaker males," slipped back out to seek the rooms of the women with whom they had arranged—or hoped to extemporize—assignations. Because of the hotel-like scale of such houses, it was prudent to carry a candle and read the name card by the honeyed glow of its light, before softly knocking.

An incident from Lady Cynthia Asquith's diary suggests how thoroughly these matings remained under female control. Lady Diana Manners was making her debut at Blenheim in 1917 and was suffering from the attentions of Lord Wimborne, the forty-four-year-old Viceroy of Ireland. Wimborne, like Winston Churchill, was a grandson of the seventh duke of Marlborough and apparently well-versed in the uses of Blenheim; he was said to have ordered the locks removed from the door to the debutante's bedroom. But a friendly duchess who considered

Blenheim "a house of ill-fame" had presented Lady Diana with a revolver "and charged her to be sure and say at tea-time, in a loud voice, that her maid always slept with her!" Asquith doesn't reveal whether this information dissuaded Lord Wimborne. But it is pleasant to think of him staring down the barrel of a revolver, obliged to sit on a corner of the girl's bed and pretend he had just dropped by for a friendly chat about the Irish question.

Clearly, there were rules of engagement in the glad round of musical beds. A woman was expected to choose a lover of suitable background, but beyond that, which lover was largely immaterial. In Nancy Mitford's *Love in a Cold Climate*, a young woman inquires about the identity of a married woman's lover, and a gentleman friend explains: "It doesn't make a pin of difference who it is. She lives, as all those sort of women do, in one little tiny group or set, and sooner or later everybody in that set becomes the lover of everybody else, so that when they change their lovers it is more like a cabinet reshuffle than a new government. Always chosen out of the same old lot, you see." The girl wonders if it is the same in France. "With society people?" her confidante replies. "Just the same all over the world." Discomfort over a partner's infidelity may well have been ameliorated, at some level, because everyone came from the same tiny group or set not just socially or economically, but genetically, too. They were members of the subspecies and could trace their intermarriages back across hundreds of years in the pages of *Burke's Peerage*. Pamela Harriman, who was herself the eldest daughter of the eleventh Baron Digby, once remarked: "They went to bed a lot with each other, but they were all cousins, so it didn't really count."

Nor was there anything particularly modern about this easy indifference to conventional morals. "Good living gave them zest; wealth gave them opportunity; and they threw themselves into their pleasures with an animal recklessness at once terrifying and exhilarating to the modern reader," the historian Lord David Cecil wrote, about the eighteenth-century aristocracy. The misbehavior of wealthy males with women of both professional and amateur standing was unsurprising. But Cecil added, "Even unmarried girls like Lady Hester Stanhope were suspected of having lovers; among married women the practice was too common to stir comment. . . . The Harley family, children of the Countess of Oxford, were known as the Harleian Miscellany on account of the variety of fathers alleged to be responsible for their existence. The Duke of Devonshire had three children by the Duchess and two by Lady Elizabeth Foster, the Duchess one by Lord Grey; and most of them were brought up together in Devonshire House, each set of children with a surname of its own." The closest analogy in the natural world to the behavior of such aristocrats would appear to be the Australian superb fairy wren where 76 percent of offspring result from extra-pair copulations.

A wife's fidelity apparently mattered mainly in the early years of the marriage, to establish the legitimacy of the heir to the family title. Or as a Blenheim historian put it, "No gentleman stalked a lady until she'd been married at least ten years; she needed a running start in order to fill her nursery with legitimate offspring, but no one cared who sired the tail end of a family." A ten-year waiting period wasn't, alas, always sufficient to ensure the patriarchal lineage. Lady Melbourne gave her husband a legitimate son, Peniston, in 1770. But Peniston died of consumption in his midthirties, and his brother William, born in 1779 and "universally supposed to be Lord Egremont's son," succeeded him as heir (and went on to become prime minister). The elder Lord Melbourne was so annoyed at being displaced from any blood connection to his own family's title and land that he peevishly allowed William less than half the allowance he had granted to Peniston.

Apart from the waiting period early in marriage, the other great rule of sexual combat seems to have been that any woman, married or not, titled or commoner, should sleep with the sovereign if he wanted her. When he was Italian ambassador to the court of King Louis XIV, Primi Visconti remarked: "There is not a single lady of quality who does not yearn to become the King's mistress. Many women, whether married or not, have told me that to be loved by one's monarch was no offense to their husbands, to their father or even to God himself . . . and the worst is that the families, the mothers, the fathers, and even certain husbands would be proud of this." As were the women themselves: One of the aristocratic mistresses to King George II, on being asked to name the father of her one-year-old son, replied that it was her husband "upon honour." Then she laughed and added, "but I will not promise whose the next shall be." In fact, it was in the woman's interest (as well as her mother's and father's) for her to be impregnated by the king, especially in the case of Louis XIV, who tended to look out for his bastards and marry them into the aristocracy.

The idea that the king's amorous interest in a woman also conferred prestige on her husband might seem like a case of mate-copying gone haywire. But having one's wife sleep with the king could be an excellent career move. When Lady Melbourne conceived yet another son during an affair with the future King George IV, for instance, she used the liaison to get her husband promoted to Lord of the Bedchamber, a title with valuable perquisites and little intended irony. Other wives raised their husbands into the aristocracy on the strength of what the Duke of Bedford snippily referred to as "services rendered—how shall I put it?—in a horizontal position." Charles II, for instance, created the barony of Castlemaine for the husband of one of his mistresses to persuade him of the wisdom of looking the other way.

In more recent times, Prince Charles and Camilla Parker Bowles were merely observing the ancient and unspoken rules of their class when they arranged suitable marriages with others and used them to conceal their own illicit romance. According to custom, Camilla apparently suspended her sexual relationship with Prince Charles during the first six years of her marriage, producing two children by her husband Andrew Parker Bowles. Then, having done her part for legitimacy, she was happy to revert to the generous tradition of her own great-grandmother, Alice Keppel, mistress to King Edward VII. Andrew Parker Bowles also played the part of *mari complaisant* as expected, at least initially.

Where Are the Children?

If the rich and socially dominant get more sex than the rest of us, or if rich *males* at least get more sex, this raises an ugly question: Why don't they also have more children? Why do people with more resources appear to have *fewer* children, contrary to almost everything we know about the natural world? Where are the elephant seals of the plutocracy, fathering 85 percent of the offspring? Where is the dominant sage grouse grabbing 80 percent of all matings? The members of the *Forbes* 400 from year to year have typically produced an average of about three children, a bit more than the norm for all Americans. But the list includes no Moulay

Ismail the Bloodthirsty, said to have left 888 children, nor even any Alexander Dumas, who made and spent a fortune, and claimed to have fathered 500 children. According to the Darwinian point of view, the ultimate aim of accumulating resources in any species is to improve an individual's reproductive success. So if Darwinism makes any sense at all, it isn't enough merely to show us the money, the resources, the cultural success; it's necessary to show us the children, too.

Darwinians are acutely conscious of the logical threat to their arguments posed by the human pattern of having fewer children as socioeconomic status improves. "The ubiquitous nature of this inverse fertility pattern—found everywhere across countries and continents—is evolutionarily puzzling," one anthropologist writes, "because it constitutes a prima facie refutation of the relationship between cultural and reproductive success." That is, if being more successful causes people to have fewer children, then successful people get fewer genes into future generations, and the traits conducive to success should thus gradually disappear from the population. (Assuming that some of these traits are in part genetically based.) Becoming rich begins to seem like a recipe for extinction, as in those aristocratic families that dwindle away to a single demented spinster and then disappear. Indeed, like the entire British aristocracy, which is on schedule, at the current rate of nonreplacement, to vanish from the Earth somewhere around the year 2175. (The figure, from David Cannadine's *The Decline and Fall of the British Aristocracy*, is based on titles continuing to go extinct at the current rate of four or five a year. With no new hereditary titles having been created since 1983, author Robert Lacey anticipated in his book *Aristocrats* that "factors affecting fertility and mating capacity will doubtless be studied as keenly as they are today in the sagas of panda survival—though, unlike the panda, the British aristocrat does not need to mate with his own kind: any female will do.")

The superficial answer to the question of why the rich don't have more children is that culture and evolution have come to a parting of the ways, and culture has triumphed. Laws, the scrutiny of the press, and other forms of cultural pressure have shifted the balance of power between alphas and their subordinates. Behaviors that may once have been conducive to reproductive success, like raping the chambermaids, can now get a chap introduced to the flip side of sexual harassment courtesy of the alpha males in Cellblock Nine. DNA tests and paternity suits can make careless sex disagreeably expensive, as Mick Jagger periodically reminds himself, most recently giving a $7 million settlement plus a $25,000-a-month allowance to the former underwear model who is mother of his latest child. And that was on top of a $36 million divorce settlement with Jerry Hall, who thought her own four children by Jagger ought really to have been sufficient, not to mention two children by first wife Bianca. (Well, maybe the rich *do* have more children. But even Mick is no Moulay Ismail.) The relatively simple business of keeping a mistress now requires the input of a human resources officer, or, better yet, a good lawyer. PaineWebber investment banker Orhan Sadik-Khan, seventy years old, with a wife and a nice place in Old Greenwich, Connecticut, woke up one day in 1998 to find himself the defendant in a palimony lawsuit by a twenty-eight-year-old Russian immigrant. The case was quickly settled, but only after the appearance of an article in *The Daily News* headlined "Russian Beauty Slams Wall St. Exec with $3.5M Sex Suit" and a two-page package in *The New York Post* headlined, "The Kinky Tycoon: Ex-Model Files Dirt-Filled Suit."

Publicity like this, whether true or false, can make even a very rich man celibate or at least a bit more careful: The essential logic of philandering only with other members of the subspecies was perfectly articulated by the wife of the wealthy British politician Alan Clark. Clark had arguably exceeded the bounds of prudent behavior by bonking the wife and two daughters of a South African magistrate. The tabloid press reported the case with glee, and Clark's wife graciously spared everyone the charade of outraged wife: "If you bed people of below-stairs class," she said, fairly sighing at the folly of it all, "they will go to the papers."

Culture and evolution have also parted company because of birth control, to which the rich have frequently had greater access than the poor. In the 1880s, for instance, possession of the primitive diaphragm known as the "Dutch cup" was a mark of privilege among women in London. Upper-class women with suitable references obtained theirs at a Mayfair bookshop, while other women were denied such devices by law. The tendency of the rich to reduce family size by artificial means is in fact far older than we might expect. No reliable contraception was available in ancient Rome. But the rich kept their families at the minimal necessary size by delaying marriage and probably also by abortion, which was accessible only to those who could afford a doctor. Polybius complained that the love of money and ostentation made men "unwilling to marry, or if they do, to raise the children born, except for one or two at most out of a larger number." Infanticide was often the fate of excess children, particularly girls. Musonius Rufus castigated the rich for "exposing children," that is, putting them out to die as a means of keeping their families small.

The real question, though, isn't how the rich reduce the size of their families, but why. The most intriguing answer comes from the anthropologist Laura Betzig, who argues that the entire purpose of monogamy among the rich is to minimize heirs and thus *perpetuate* polygyny. Concentrated in one or two offspring, preferably sons, wealth becomes the means by which generation after generation of happy scions pay obeisance to family but still spread their seed with every chambermaid and good-time girl in groping distance. Even more shockingly, Betzig suggests that, in the strictest Darwinian terms, this ultimately works to the benefit of the wronged wives, the Ann Rorks, Ann Gettys, Jerry Halls, and Ivana Trumps of the world.

This line of thinking goes to the entire character of family life among the rich, and so belongs in the next chapter.

11

Family Business

Caught in the Perpetual Dynasty-Making Machine

Unlike the male codfish, which, suddenly finding itself the parent
of three million five hundred thousand little codfish, cheerfully
resolves to love them all, the British aristocracy is apt to look with
a jaundiced eye on its younger sons.

—P. G. WODEHOUSE

ALEXANDER THYNN, SEVENTH MARQUESS OF BATH, IS A WELL-
known British madman, a flamboyant character caught somewhere
between the 1660s and the 1960s and not quite ready to let either one go.
He has wild gray hair swept round to the side, pirate-style, and braided in
back into a short ponytail. His eyes are big and blue, with a quality of
continual delight, and, at seventy, he still has a lumbering boyishness in
his stride. He was dressed in an ornate Indian jacket over a red vest and a
yellow plaid shirt when I visited him at Longleat, his family's 118-room
house in the Warwickshire countryside. He had workers busy painting
the Elizabethan plasterwork in his private apartments purple, as if it were
Haight-Ashbury in the summer of love.

Lord Bath, who is listed not just in *Burke's Peerage* but also in that
modern handbook of status, the *Sunday Times* "Rich List" of Britain's
1,000 richest people, gets his income largely from the leisure industry:

Longleat is open to the public and attracts more than 400,000 visitors a year. It's a handsome 450-year-old block of a building, pale limestone, topped like a wedding cake with finials and urns and statuary. The estate around it is another grassy savanna, punctuated with branchy trees and fat sheep, plus the ritual safari park, celebrated for "the lions of Longleat." Lord Bath owns more than 10,000 acres of land hereabouts. Like other rich people he prefers to think of himself as more than his title or his net worth, and he has worked over the years as both an artist and an author, with admirable productivity undaunted by the absence of talent on either count.

Almost the first thing after greeting me, Lord Bath showed me the scrapbook in which he glues press clippings about his life, including a tabloid headline, "Lord Bath says, 'I let the dog gnaw on my bongo.' " Bongo apparently being British slang for an instrument of a sort, not a drum. I smiled wanly at the dog sitting attentively at my right knee but did not pat him. Lord Bath was actually more interested in another clipping about his family's long and colorful history of sexual adventure. It described an episode from 1869 when his great-grandmother, a grey-eyed beauty named Lady Mordaunt, fell out with her husband, who spent too much time fishing. The child to whom Lady Mordaunt had just given birth was temporarily blind, and in a panic that the cause might be "the complaint," meaning venereal disease, she blurted out to Lord Mordaunt that he was not, in fact, the father. Not only had she been "very wicked," she admitted, but "with more than one person" and, revealing (along with much else) a family knack for excess, "in open day." Edward Prince of Wales had apparently been her chief companion in field sports. Her husband had once caught her showing off her carriage-driving skills to the prince and was so enraged that he had her prize ponies shot before her eyes, a case of extreme mate-guarding behavior by proxy. Better not to be descended from such a brute. Better still, said Lord Bath, "I think I might be descended from Edward VII." Also a brute—but royal.

Indeed, the Thynns seemed to be descended from every famous person who ever got close enough to blow a kiss at the British Isles, and even a hippie like the present Lord Bath pays homage to this familial heritage, in his fashion. He led me on a tour of his private apartments, where he has hung his own portraits, in sawdust and oil, of more than 150 ancestors, some of them framed with purple neon lights. "The earliest one is Tacitus, in A.D. 55," he said. "In the corner there." He went bounding past Charlemagne, Harold, Alfred, and, on the men's room wall, Ethelred the Unready. "That crowd is largely Pembrokes and Borontsovs—they're Russian," he said, "and this lot are Norfolks and Marlboroughs. And there. . . . Well, it's not worth telling them all."

The family heritage also had a flip side of children born, as the British say, on the wrong side of the blanket. If one listened carefully to the grand pageant of British history, the sounds one could hear in the background were the openings and closings of bedroom doors behind which Thynns were behaving badly. When Sarah Jennings strategically resisted the advances of John Churchill, for instance, she was inspired in part by the example of another young girl at court who'd been reduced to tears and pregnancy by Sir Thomas Thynne of Longleat.

The present Thynn, as Lord Bath's family now spells its name, celebrated this heritage, too, both on gesso and in the flesh. Not content merely to misbehave in the grand setting of the private apartments, as his family has always done, Lord Bath has diligently covered room after room with gaudy murals depicting scenes from his own life and from the *Kama Sutra*, often overlapping. One scene, "My First Kiss," was captioned, "I'm afraid I don't like other people's saliva." It was evidently an acquired taste, and the height of Lord Bath's artistic achievement was reserved for the spiral staircase up to the master bedroom. The three-dimensional portraits here represented, as he put it, "the severed heads of every woman I have ever loved," identified in the aristocratic manner not by name but by number, one to sixty-eight. (Since my visit, the total has risen to seventy-three.) "Rather than notches on a bedpost," he said, "as one goes up to bed, one can kiss them."

Lord Bath, who refers to himself in his giddier moments as "the loins of Longleat," has claimed in print that "to some extent I pioneered polygamy in this country." (One of the happier cases of those who are ignorant of history being doomed to repeat it.) Monogamy caused "the rebel" to start "kicking inside me," he told one newspaper, and he needed "wifelets" to calm it back down. "Ideally I should have been able to gather together a group of, say, five women, have a large number of children and then see if everyone learned how to live together in a spirit of mutual respect and understanding."

There was, of course, nothing in the least rebellious about this, except for talking about it out loud. (Well, and also that bit about mutual respect and understanding.) What Lord Bath had done was in truth the completely traditional thing for a man of his wealth and position. Like Gordon Getty, he was still married to his first wife, even if they seldom passed a night under the same roof. His wifelets to the contrary, he had fathered only two legitimate offspring. Through the magic of primogeniture, the younger child, the only son of this marriage, was destined to inherit his title, Longleat, and all the privileges that go with it.

Why Monogamy?

According to anthropologist Laura Betzig, this is exactly how monogamy among the rich was intended to work.

The logic of reduced reproduction, or, to be more precise, reduced *legitimate* reproduction, among the rich is straightforward: "There is an irresistible and automatic method whereby the wealth of the great plutocrats regresses to the mean," one historian has written. "It is called having children." The dread of this regression is a topic one hears endlessly repeated among the rich: After generations toiling upward from minimum wage, a family's great hope finally arrives at the top. Then, in a burst of reproductive exuberance, he does the natural thing for a founding father and begets all the children he can possibly afford. After a little while, they have children, too, and pretty soon everyone is making minimum wage again. The technical term for this is the Malthusian effect, after the British economist Thomas Malthus, who described the tendency of populations to increase faster than the food supply, or money, needed to support them. It's what rich folk mean when they fret about going "from shirtsleeves to shirtsleeves in three generations," or in the charming Italian variant "*dalle stalle alle stelle alle stalle*"—roughly, from mucking out stalls to swinging on the stars and back again. One, two, three.

By having fewer legitimate children, a wealthy couple can at least delay this dreaded regression. The marriages of John D. Rockefeller and Henry Ford, for example, each produced only a single male heir—and thus two of the most enduring dynasties of our day. (John D. also had three daughters, but his son received the bulk of his fortune.) Isaac Merritt Singer, on the other hand, made a worldwide commercial success out of one of the great inventions of the nineteenth-century, the sewing machine. But he also acknowledged fathering twenty-seven children. The family has not entirely subsided back into the ranks of the wage slaves (a Singer heiress has bankrolled the *New Republic* in recent decades), but any hope of a Singer dynasty vanished at the (pro)creation.

Monogamy helps secure the future of the family dynasty by concentrating a family's wealth in the smallest practical number of legitimate offspring. In the view of Betzig and some other researchers, this frees wealthy males to achieve reproductive success in contexts where it traditionally has not counted—that is, by having illegitimate children. "Simply put," the anthropologists Hillard Kaplan and Kim Hill write, "rich men may have been practicing a dual investment strategy; having few legitimate offspring to whom they passed down their wealth (and hence their mating advantage) and some number of illegitimate offspring in whom they invested little (and who therefore did not deplete their resources)." This is of course a highly arguable theory. Some men, like Rockefeller and Ford, seem to pursue only the first half of the strategy. Some, like Larry Hillblom of DHL, never marry and stick to the second half. Still others do neither. Producer David Geffen, number 47 on the 2001 *Forbes* 400 list, is gay and childless. Texas wildcatter George P. Mitchell, number 136 on the list and its champion of legitimate begettery, has ten kids by his only wife. Even the writers who proposed the dual strategy would concede that it is no more than a propensity.

The Revenge of the Wronged Wife

A much more intriguing issue is Laura Betzig's suggestion that the wives in dual-strategy marriages ultimately benefit from the propensity to infidelity among wealthy males. Let's say you're Miss All Things Bright and Beautiful of the year A.D. 1650. From the Darwinian viewpoint, such a society gives you only two or three basic options for maximizing your reproductive potential. You can marry a man of limited means, be penniless, and give birth to as many as twenty children, knowing most will die young. You can also have an affair with the local chieftain—let us call him Lord Bath—and perhaps slip one of his children into your brood, hoping that the liaison will win your family a better hog trough, say. Or—and this is clearly the jackpot—you can marry Lord Bath, who controls most of the resources hereabouts. You still get no more than twenty children, and in all likelihood far fewer. But if you marry Lord Bath, you also get the wherewithal to raise your children in the healthiest available circumstances. (Even in modern American society, it's a mistake to ignore the effect of wealth on child survival. When biologist Susan Essock-Vitale scrutinized reproductive success among the 1982 *Forbes* 400, she looked, among other factors, at how many offspring survived to maturity. The study estimated that 100 women "reproducing at the rate of the *Forbes*' sample" would give birth to 310 children—and at a survivorship rate of 99 percent this would yield 307 adult offspring. Their counterparts in the general U.S. population would have only 240 children—and survivorship of 93 percent would reduce this to just 223 adult offspring. "The wealthy sample," she concluded, "would be reproducing at a 38 percent greater rate than white and 20 percent greater rate than nonwhite segments each generation.")

For Miss All Things Bright and Beautiful, the down side of her marriage is that Lord Bath happens, in the manner of the rich, to be littering the countryside with scores or even hundreds of his bastard children. But according to the strict rules of legitimacy and primogeniture, only Lady Bath's first-born son inherits the family wealth. Here is the key: This means her son will have the resources to continue philandering—and thus give her scores or even hundreds of grandchildren.

It's Not about Having Sex with Your Mom

Not a pretty picture, is it? Mother-son alliances may be more common among the rich not because of any Oedipal sexual attraction, the sociobiologist John Hartung has argued, but because women want the spending and the philandering to be done by their sons rather than their husbands. Ann Rork had every reason to deplore J. Paul Getty's promiscuity. But the philandering of their son Gordon has nearly doubled her reproductive success, giving her seven known grandchildren. The current Mrs. Getty likewise has every reason to resist Gordon's philandering and fight to direct the family's resources to her four sons, who represent her reproductive potential, both legitimate and otherwise. In *The Edwardians*, Vita Sackville-West depicted the young scion, a duke named Sebastian, as little more than the reproductive instrument of his parents and all the generations before them: "What were the wild oats of such a young man? An inevitable crop, sown by his bad godmother at his christening. Not sown even by his own hand, but anticipated on his behalf. Poor Sebastian, his traditions were not only inherited, they were prophetic." Sackville-West's choice of a godmother, a female, as the architect of a son's philandering is appropriate: At least until the advent of modern paternity testing, a mother's genetic interest in her child has always been far more certain than the putative father's.

The picture may be even less pretty still. The focus on the son isn't simply a product of the archaic system of primogeniture. It may also reflect an underlying bias of wealthy families against daughters. The tendency is to regard primogeniture, in which the first-born son inherits everything, as merely a practical way to keep an estate from being endlessly subdivided in the days when a family's status depended on its landholdings. In the modern system based on free-floating capital, primogeniture is unnecessary and it has largely disappeared, except among throwbacks and in certain special situations. (Malcolm Forbes, for instance, left a 51 percent share in his magazine company to his eldest son.)

But leading families in the natural world also often display a bias against daughters. High-ranking red deer and spider monkeys skew their reproduction, producing more sons and fewer daughters. So do opossums when they are well-fed. If you starve the same possums, they produce more daughters. The mechanism is unclear. Perhaps male fetuses are more prone to spontaneous abortion under the stress of starvation. The result is the same: prosperous families have more sons. Among humans, studies have shown this to be true in selected populations of aristocrats, royalty, and early American settlers. Likewise, U.S. presidents up through George W. Bush have had ninety sons and sixty daughters. The logic, according to the theory proposed by biologists Robert Trivers and Dan

Willard, is that prosperous males have the diamonds, dead insects, or other resources to attract multiple females.

In poor families, on the other hand, sons generally lack the resources to be much of a catch for the long haul, though they may enjoy a season or two as sneaker males. So the opposite strategy is likely to produce the greater benefit: A daughter has a better chance to improve the family's prospects by latching onto a male with diamonds. The most successful instance of this strategy on record was a tanner's daughter in Normandy who had a dalliance with a duke, producing the son called William the Bastard, later William the Conqueror, from whom springs the entire British royal lineage.

No one is suggesting that the tanner and his wife had this all worked out in their upwardly mobile family plan. Nor, in all likelihood, did Miss All Things Bright and Beautiful ever calculate the effect that marrying Lord Bath might have on her reproductive potential. More likely she was just captivated by the way his forearm flexed when he was cleaving people in two, or maybe her parents made her do it. The point evolutionary psychologists make is that we tend to behave in ways that improve our reproductive potential even when this is the furthest thing from our minds.

It's about the Inheritance, Stupid

In the natural world, inheritance is rudimentary, at best. Animals may pass on territory to their offspring, and dominant mothers may kibitz at playtime to give their kids an edge over the neighbors. Early humans presumably had the same genetic propensities to look out for their kin and keep up family status. But human cultures have adapted and elaborated on these propensities in spectacular fashion. The result is a system of inheritance that allows wealthy people to give their children, and even their children's children's children, parental care from birth to grave.

For the rich, all too often, family is thus less about love than about inheritance. Such behaviors as delayed reproduction, mate-guarding, inbreeding, birth control, infanticide (particularly of daughters), primogeniture, and generation-skipping trusts have all served in various times and places as the Darwinian tools of dynasty building. Mate-guarding and the sexual double standard, for instance, aren't simply about fragile male ego. They're about inheritance: Minimizing the number of heirs naturally maximizes Lord Bath's anxiety that these heirs be his own off-spring, and not the product of adultery. The concern for inheritance is so paramount that rich men in some cultures have devised ways to practice mate-guarding even when they are dead. Romans perpetuated a cult of the faithful widow, the *univara*, forever devoted to her late husband (the Ethel Kennedy course, as opposed to the more entrepreneurial path taken by Jackie Kennedy). Other cultures obliged a widow to retire to the convent or to marry her dead husband's brother, so his inheritance would at least end up with his nearest genetic kin in the next generation. The Darwinian nightmare for rich men is to suffer the fate of Nicholas van Rensselaer, who died childless in 1678, possibly poisoned by his wife Alida or her lover. Alida soon remarried and used her former husband's wealth to launch one of the great American dynasties, the Livingstons of New York. The Livingstons have tended ever since to be careful about whom they marry.

A Good Family

Devotion to family is one of the more endearing adaptations of the rich. We have been blessed with six generations of Marshall Fields, four Pierre S. du Ponts, four Anheuser Buschs, three Henry J. Heinzes, and the Ford named Henry II, who liked to look in the shaving mirror and declare, "I am the king. The king can do no wrong." Prizefighter George Foreman has cleverly shortcut the process by naming all four of his sons George, the next best thing to cloning.

An optimist might say the rich simply have the leisure to enjoy a warmer, more loving family life. And the ecological perspective is, for once, also remarkably optimistic: Primates are strongly attracted to kin and also strongly attracted to individuals of high status, and the gravitational pull of these two forces ought, in the word's of one primatologist, to be "additive for high-ranking families and counteractive for low-ranking families." Hence the ancestor worship, detailed genealogies, reunions, and other tribal rituals of the rich. In one East Coast family, the genealogy is actually coded by number to indicate each individual's degree of kinship to the founding father—and thus his or her share of the family fortune. In times of disagreement one of them will gently remind another, "Don't forget, I'm number six."

It's seldom possible to put a cash value on family affiliation, particularly since the beneficiaries prefer to pretend it doesn't exist at all. Donald Trump derides people who inherited their wealth as members of "the lucky sperm club," implying that he himself fought his way up from the gutter. (In fact, his father's estate was valued at more than $150 million.) And when she published a book, *Bright Young Things*, about her wealthy New York pals, socialite Brooke de Ocampo airily advised *Vanity Fair*, "Don't call them aristocrats, they're meritocrats." Then came the roll call: "Eliza Reed Bolen, who works for her stepfather, Oscar de la Renta; Aerin Lauder . . . and her sister, Jane, both powerhouses at Estée Lauder; . . . Alexandra von Fürstenberg, the inspiration behind the return of her mother-in-law Diane's wrap dress." They were meritocrats, that is, more or less the way Prince Charles is a meritocrat, in the family business.

Daddy Warbucks to Watch over Me

Denying the importance of family influence is understandable. One of the most debilitating things about growing up rich is that the rich themselves never know for sure what family name and fortune have obtained for them, and what they've accomplished on their own. John D. Rockefeller, Jr., once complained that "even the girls in the office . . . can prove to themselves their commercial worth. I envy anybody who can do that." No matter what the children of the rich accomplish, moreover, outsiders always assume that someone somewhere pulled some strings. In our hearts we give the rich credit for nothing, except their money. Telecommunications entrepreneur Richard Li is so touchy about being known as the son of Hong Kong's most successful tycoon, billionaire Li Ka-shing, that when he talks about "the 'f' word," he means "father." But family influence is difficult to overlook. Li's father staked him the $125 million to start his first company, Star TV, which Li quickly sold to Rupert Murdoch for $950 million. When Li later challenged Murdoch in the bidding for Hong Kong Telecom, one of the biggest phone companies in Southeast Asia, family connections helped him arrange an almost instantaneous $13 billion bridge loan from a four-bank consortium in Beijing. More recently, when his Internet company, Pacific Century Cyberworks, was sinking fast, Li was able to wangle lunch with the most important businessman in Hong Kong, the unspeakable "f," in the public dining room of the Shangri-La Hotel. According to *Fortune*, the stock market took the message that father would stand by son: With Pacific Century stock opening that day at 5.375 and closing at 5.6875, lunch was worth about $671 million—and the company continued to gain for the next six days.

The rich may be genuinely oblivious to how large a role family reputation plays in their lives, until it disappears. One of Joanie Bronfman's informants, a woman from a famous family who had recently married, went to a social function and recalls: "There were people there from all different social classes. . . . I introduced myself by my married name and got involved in this conversation with this other woman whose husband was on the way up. At some point she said, 'Oh, your tea has gotten cold. Why don't we have a waiter come and warm it up?' And I said, 'Oh, no, it's really fine.' And she said, 'Oh, no, it's cold. Let's send it back and get some more.' And I said, 'Oh, I don't have very refined tastes.' And her face just changed, 'Oh, I didn't mean to imply that.' All of a sudden she was taking care of me. It was hilarious and I looked at her: *You really think that I am needing to be taken care of like this poor little orphan Annie type?* And I realized, at that moment, what the name had done. It had been like a great big person standing behind me wherever I was. I knew people would assume that I knew what I was doing because of my name. If I went some place and I was dressed more casually than everyone else, then they were overdressed. If it was fancier, then they were underdressed. I just didn't have to worry about that kind of thing." Now suddenly she found she had to be careful not to use the salad fork for dinner. Small wonder that the current lieutenant governor of Maryland likes to be known as Kathleen Kennedy Townsend, a reminder that she is Robert F. Kennedy's daughter, or that Gordon Getty's illegitimate daughters wanted to change their name to Getty.

Kissing Cousins

A passionate concern for family (or at least for the family's estate) encourages close scrutiny of anyone seeking admission to this hierarchy by way of the bedroom. Old money types used to refer to the *Social Register* as their "stud book," in the belief that suitable matches would yield thoroughbreds. It was social Darwinism in the bedroom, and it made Stephen Birmingham, hagiographer of the rich, pop the nib on his fountain pen: "Breeding! What is it? How does one define it? Is it character? Is it soul? Is it the glue of generations, the olive branch of peace? Is it what lies at the very heart and core of civilization—even immortality, perhaps?" But breeding as practiced by the rich has often meant inbreeding, and from the perspective of evolutionary psychology, the motive was somewhat more selfish and insecure than the greater glory of civilization.

For the rich, inbreeding provided a means for minimizing the number of legitimate descendants, thus concentrating family wealth and ensuring that the fiefdom didn't get divvied up to the point of insignificance. At the same time, it may also have helped to maximize the genetic influence of the founding father, for whom it could also serve as a posthumous substitute for mate-guarding: Even if some future scion got cuckolded, his heirs would still be the founding father's genetic descendants, by way of the cousin-wife. But given our built-in inhibitions against incest, how common could inbreeding really be?

Charles Darwin himself, who came from a wealthy family, married a first cousin. Likewise, the present Sultan of Brunei is the child of first cousins and is also married to a first cousin. One of America's greatest dynastic families, the du Ponts, also made it their practice, in the words of one historian, "not to waste their time (or their sperm, for that matter) outside the family." The founding father, Pierre S. du Pont, advised his heirs to intermarry as a way to ensure "honesty of soul and purity of blood." One of his granddaughters married one of his grandsons, and in the following century, other du Pont heirs married their cousins another thirteen times, until the family began to worry about producing genetic misfits. Even now, the du Ponts have a sort of tribal visiting ritual each New Year's Day, in which the female kin assemble at various estates around the Brandywine Valley and the men travel from house to house to greet and size them up.

Girlfriends, fiancées, and other outsiders stay upstairs, safely out of sight (and even after marriage they are still sometimes labeled "out-laws"). A du Pont cousin marriage took place as recently as the 1970s, though marrying within the family tends now to be discouraged.

In Europe, the Rothschilds pursued inbreeding even more systematically. Mayer Rothschild, who rose from the Frankfurt ghetto to establish the greatest banking house of the nineteenth-century, had five sons and five daughters. He shrewdly dispersed the sons to establish interlinked lending houses in the great financial capitals of Europe. Yet he wrote his daughters out of his will with remarkable vehemence: "None of my daughters and their heirs therefore has any right or claim on the said firm and I would never be able to forgive a child of mine who, against this my paternal will, allowed themselves to disturb my sons in the peaceful possession of their business." Subsequent generations perpetuated this harsh practice of weeding out female descendants from any role in the family business. At the same time, they assiduously brought them back into the Rothschild clan through intermarriage.

"The first and most important reason for the strategy of intermarriage was precisely to prevent the five houses drifting apart," Niall Ferguson writes in *The House of Rothschild*. "Related to this was a desire to ensure that outsiders did not acquire a share in the five brothers' immense fortune. . . . But, mercenary considerations aside, there was also a genuine social difficulty in finding suitable partners outside the family. By the mid-1820s the Rothschilds were so immensely rich that they had left other families with similar origins far behind." The problem, as Ferguson goes on to make clear, was in finding suitable partners for the daughters, and he concludes, "By this time only a Rothschild would really do for a Rothschild."

The Problem of Daughters

Like so many other rich families, the Rothschilds had run up against the single biggest built-in hazard of hypergyny, hypergyny being the biological term for the female propensity to marry upward: At each level as a family ascends the economic pyramid, there are fewer likely men higher up the pyramid for the women of the family to catch. At the very top, they can do no better than marry their cousins—or even their brothers, as happened with Caligula of Rome, Ramses II of Egypt, and various Incan, Hawaiian, Thai, and African monarchs.

If inbreeding sounds unattractive, the alternatives for unsuccessfully hypergynous daughters were often worse: In India, the most notorious example, a high-caste family could marry a daughter up the social scale only by impoverishing itself to give her prospective husband a huge dowry. By killing infant daughters and investing in one or two sons, on the other hand, the same family could bring in dowries and also ultimately produce more descendants. When Charles Allen was compiling his oral history *Lives of the Indian Princes* in the early 1980s, he interviewed a granddaughter of the rajah of Jaunpur near Benares. News of her birth had at first caused general rejoicing: "But when it was announced that I was a girl there was great disappointment. However, I am the first female child in the family, after two hundred years, who was allowed to survive."

Westerners should not be overly shocked: Demographic evidence indicates that female infanticide was also common among the gentry in some medieval European cultures. The aristocratic fervor for getting daughters to the nunnery served the same purpose—that is, to get rid of excess daughters and concentrate inheritance on the male line. Moreover, the underlying prejudice of the rich against daughters may persist even now in subtler form. Wealthy American mothers intent on avoiding gender bias are generally scrupulous about not killing their daughters. But a 1991 study indicated that they are far more likely to nurse sons than daughters and to nurse them longer before weaning. As evolutionary theory would predict, low-income mothers did just the opposite, favoring daughters over sons.

All in the Family

Among the Rothschilds, Mayer's youngest son James started the family practice of inbreeding in 1824 by marrying his brother's daughter. Over the next sixteen years, four of Mayer's other granddaughters chose to marry his grandsons, their own first cousins. The family tree quickly began to collapse on itself. In 1876, for instance, Albert von Rothschild married Bettina Rothschild. Had their offspring looked back over their genealogy, they would have found that of the sixteen slots for great-great grandparents, three slots were occupied by Mayer Rothschild, three by his wife Gutle, and three more by Mayer and Gutle's sons. Mayer and Gutle thus also turn up three times apiece in the previous generation, as great-great-great grandparents. The sixty-two slots on the family tree over the previous five generations were filled by just thirty-six people, and twenty-four of them were born Rothschilds. While this was an impressive start, one must add that the Rothschilds's efforts pale by comparison with the long-term inbreeding among European royalty: Robert C. Gunderson, a Utah genealogist, has been working since the late 1960s on the lineage of Charles Prince of Wales. Because of intermarriage, Prince Charles has no more than twenty-three thousand ancestors in the seventeenth generation back, rather than the theoretical maximum of more than sixty-five thousand. A single person, Edward III, King of England from 1327 to 1377, has so far turned up two thousand times in his ancestry, and Gunderson estimates that Edward would occupy about twelve thousand slots if the full family tree were known.

Biologists analyze complex relationships like those in the Rothschild family in terms of the "inbreeding coefficient," a measure of just how closely individuals are related. In the case of Bettina and Albert's seven children, the inbreeding coefficient was .064, or 256 times greater than for a comparable family with normal outbreeding over five generations. In other words, sixty years after Mayer Rothschild's death, he was the source of both alleles at 6 percent of the gene loci in his great-great-great grandchildren. This is equivalent to the inbreeding coefficient when first cousins mate. Conventional wisdom says this level of inbreeding must have been hazardous to the health of the Rothschilds, yet Niall Ferguson reports no evidence of genetic abnormalities. Six of Mayer's forty-four great-grandchildren (Albert and Bettina's generation) died before the age of five, but Ferguson concludes that this was about normal for Western Europe then; whether it was normal for their class is unknown. Of Albert and Bettina's children, one died in infancy and one in his teens, but five became adults and four of them married. "Of course, the alternative possibility exists," Ferguson writes, "that there was a Rothschild 'gene for financial acumen,' which intermarriage somehow helped to perpetuate. Perhaps it was that which made the Rothschilds truly exceptional." But this smacks of eugenics and he quickly backs away from the idea as "unlikely." Apparently confusing intention with evolution, he adds the irrelevant detail that "even if it was the case, those concerned knew nothing of it."

The Founder Effect

As it happens, though, there is a highly successful model in the natural world for the family strategy of isolationism and inbreeding practiced by the du Ponts, the Rothschilds, the Windsors, the British peerage in general, and other wealthy families. And the strategists in question also apparently know nothing about the nature of genetic inheritance: House mouse colonies tend to be dominated by a founding father, generally a tyrant. He mates with his daughters and his granddaughters, and his genes spread rapidly through the colony, a phenomenon biologists call "the founder effect." Excess sons and daughters get crowded out of the family homestead and must go off to conquer new worlds, like excess sons of the medieval aristocracy being shipped out to the crusades. Some of these mousy outcasts make war, others make love. So the founder is actually pursuing a dual strategy of outbreeding abroad and inbreeding at home, much like a rich man with mistresses abroad and a wife (who is perhaps his cousin) at home. The mouse colony itself, after generations of intense inbreeding, becomes just a little different from neighboring colonies, a little odd, possibly a little better adapted to its circumstances.

Or possibly, given the genetic hazards of inbreeding, a little worse. As anyone who has ever attempted selective breeding of animals will attest, this sort of business is notoriously unpredictable. In the case of the Rothschilds, inbreeding may perhaps have increased the frequency of some sort of "gene for financial acumen." But inbreeding, family culture, or a felicitous combination of the two, also appears to have encouraged a genius for entomology, particularly the study of fleas. As a result, 153 species or subspecies of insect now have "Rothschild" incorporated in their scientific names, and Miriam Rothschild, great-great-great grandaughter of Mayer, is rightly celebrated, among other achievements, as the author of the book

Fleas, Flukes & Cuckoos. Over the long term, family planning begins to seem like an oxymoron.

Assortative Mating

Few wealthy families take the practice of inbreeding quite so far as the house mouse or, for that matter, the Rothschilds. But Nelson Aldrich has written that "the best way to describe Old Money families is to call them families of cousins." Livingstons, Jays, Beekmans, and Astors have intermarried for generations. Franklin Roosevelt, distantly related to the Astors, was not only Teddy Roosevelt's cousin by birth but also his nephew by marriage.

Even when they are not inbreeding, the rich typically practice assortative mating. Biologists define this as "a tendency for males of a certain kind to breed with females of a certain kind," rather than at random, and it appears to be a trait of humans generally. Contrary to the popular idea that opposites attract, people usually marry people like themselves, in terms of intelligence, hair color, socioeconomic status, and—love being an odd business—even possibly earlobe length.

More to the point, rich people generally marry other rich people. Much of the child-rearing behavior of the rich seems, in truth, to be aimed at achieving assortative mating. Getting children into the right preschool or the best dancing class helps to ensure that they never actually meet unsuitable strangers. ("Diversity" is the buzzword of the moment, but it typically gets encountered in a highly controlled setting, and with the implicit message that learning to deal with the less fortunate is an important educational experience. From this point of view, the doorman's son attends the prep school on scholarship as a specimen for the edification of his social betters. As in the play *Six Degrees of Separation*, a certain horror persists that some unsuitably diverse school chum might actually show up for dinner.)

Driving the right car, going to the right vacation spots, dropping the right social references, and wearing the right clothes all help to establish the child's suitability. Moreover, rich children tend to be insulated by the carefully cultivated sense that they are different from children who do not share these badges of status. One of Joanie Bronfman's informants recalled a cousin who couldn't quite grasp *how* she was different, so "she went around the playground trying to smell the other children to see if she could tell what the difference was." Another recalled being "carefully told not to think that [she] was better, in such a way which communicated that only the better people thought they weren't better."

Assortative mating doesn't occur just within the family or the social set. Familiarity and strategic advantage also encourage the rich to marry other rich people in their own line of business. William Clay Ford, Jr., the chairman of the Ford Motor Company on the eve of its hundredth anniversary, is a great-grandson of founder Henry Ford and also of tire manufacturer Harvey Firestone. U.S. Vice President Nelson Aldrich Rockefeller was the grandson not just of John D. Rockefeller, Sr., but of U.S. Senator Nelson Aldrich, sometimes known as "the senator from Rockefeller." At the Wall Street firm of Goldman, Sachs, all partners during the firm's first half-century were members of the intermarried Goldman and Sachs families. And Brad Pitt is married to Jennifer Aniston.

Assortative mating in the natural world is poorly understood, but biologists point out that mating with individuals as much like oneself as possible increases reproductive isolation of a population and decreases the exchange of genes with other populations. In other words, assortative mating may be the basic starter kit for speciation. Hanging out with the right crowd may be how new species and subspecies get their start. And if we are talking not about genes but about memes—Richard Dawkins's term for ideas that propagate themselves around the planet—assortative mating is certainly the essence of a cultural subspecies like the rich.

The Most Successful Animal Hierarchy Ever

Few human groups have employed inbreeding, assortative mating, and other tools of dynasty-making more effectively than the British aristocracy, and none can match its record for wielding wealth and power over generations. Not only did a few hundred intermarried families own 80 percent of the land in the British isles, but at their prime in the mid–nineteenth century, they also shipped out excess sons to the outermost points of the Empire and thus extended their dominion over 400 million people and 25 percent of the land surface of the Earth. In terms of sheer territorial sprawl, the British peerage was the single most successful animal hierarchy in the history of the planet. (Perhaps it still is. The peerage owned an estimated 4 million acres of the United Kingdom as recently as 1982. And in 1999, eleven dukes and thirty-seven other aristocrats turned up on the *Times* "Rich List," mostly on account of landholdings.)

This is surely why the rich so often attempt to emulate the British aristocracy even now, deep into its decline. It's a measure of the persistence of this fascination that the self-made rich often pay serious money for amusing but otherwise worthless feudal land titles, like the lordship of Wooton Wawen Priory or the Barony of Knockninny. Peter Norton, developer of Norton Utilities computer software, paid $190,000 for the lordship of the Manor of Stratford-on-Avon. And Ole Georg, who composed music for "Baywatch," is now also Lord of the Manor of Sheldowne in Sussex.

People still also marry into the peerage, though not quite so avidly as during the heyday of the Gilded Age, when American robber barons sought status by marrying off five hundred of their daughters to cash-starved scions of European aristocracy, at an estimated dowry cost of $220 million (not counting inflation). The dynamic of the relationship has shifted a bit since then. The actress Jamie Lee Curtis, for instance, is married to Christopher Haden-Guest, the fifth Baron Haden-Guest of Saling. It isn't, however, a conventional amalgamation of titled European aristocrat and Hollywood glamour, but one in which the baron is best known for playing heavy-metal guitarist Nigel Tufnel in the movie *Spinal Tap* and for his affecting rendition of the redneck North Carolina bloodhound-lover Harlan Pepper in *Best in Show*. Curtis, meanwhile, is second-generation girl-next-door Hollywood royalty, with a covert double-barrelled name to remind us that mom and dad are Janet Leigh and Tony Curtis. (In Hollywood as on Cadogan Square, family counts.) Yet Curtis still manages to sound rather like Consuelo Vanderbilt, achieving a perfect blend of aristocratic status-seeking and offhand self-disparagement, when she remarks, "I don't book tables as Lord and Lady Haden-Guest. I don't have an account at Harrods as Lady Haden-Guest. Only when I go to the House of Lords, that's what they call me." Silly lords.

Noble Kinsmen

So why should the peerage remain so fascinating, given that genuine peers now turn up as deli workers, policemen, and bus conductors (family motto "Late, but in Earnest")? How did this group once achieve world dominion? What was the secret of "the three or four hundred families," as Winston Churchill put it, "which for three or four hundred years guided the fortunes of the nation from a small, struggling community to the headship of a vast and still unconquered Empire"? Their power derived, as we have seen, from control of land at all costs. It depended on ruthless limitation of inheritance and the shrewd dispersal of colonizing younger sons. It required the big house as a forum for courting one another and as a facade for intimidating the outside world. Above all else, what sustained the peerage for centuries was the sense of itself as a group, the lives of its members deeply interwoven and distinctly separate from the rest of humanity. "They possessed, in short, a collective awareness of inherited and unworked-for superiority," the historian David Cannadine writes. "The British patricians were highly conscious of themselves, their families and their order *in time*. More than any other class, they knew where they had come from. . . . The walls of their houses were adorned with ancestral paintings; the pages of Burke and Debrett catalogued and chronicled their forebears; their homes were usually in the style of an earlier period."

Culturally, they were a subspecies, and they implicitly acknowledged as much in the House of Lords, where in 1999 they still addressed one another as "my noble kinsman," and at the same time referred to the House of Commons simply as "another place" rather than sully themselves by giving a name to its sordid existence. They were very nearly a subspecies biologically, too, because of their long history of intermarriage, and they shared an extraordinary sense of kinship. Cannadine points out that nine members of the Cecil family were sitting in the House of Lords in the late 1930s, and even in that other place, there was a cousinship of 145 family-related Tory members of Parliament, a number

Cannadine says would have been higher in any previous generation. Yet they were never a closed breeding group. They maintained their vigor by regularly accepting new families into the peerage, however disdainfully, and by assortatively mating with Vanderbilts and Rothschilds, usually on the basis of the single criterion of money. Or as Simon Winchester put it, in his history, *Their Noble Lordships*, "The peers have been shrewd enough to do what budgerigar breeders recommend for improving the quality of that avian breed: they 'dip into the green' every so often, by marrying rich American heiresses."

Attila's Heiress

One afternoon over butter-soaked tea cakes in a wood-paneled dining room at the House of Lords, I met with the Baroness Strange to discuss the imminent demise of her social class. It was a bit like talking about guillotines, over brioche, in revolutionary France. Strange and almost all of Britain's hereditary peerage were about to be evicted from the upper house of Parliament, which their families had occupied by birthright, some of them for more than 800 years. It was to be a bloodless revolution. The peers themselves were debating the terms of their exit that afternoon, in characteristically genteel language. Also characteristically, they were still tending to the nation's business, discussing, among other issues, the need to ensure humane culling of badgers.

"You Americans look at everything horizontally, in terms of space," Strange was saying, between sips of China tea. "We look at things vertically, in terms of time. We talk about what happened 100 years ago or 1,000 years ago, in a sort of chronological order." She proceeded to demonstrate by way of her own family history. It was chockablock with beheadings, noble last words, and bewildering biblical transitions like, "There were three sisters and they were the descendants of the fourth duke." It stretched back 372 years in which the Stranges have sat on the red leather benches in the House of Lords, and from there it branched out into the distant recesses of European history. There was a certain numinous quality to this heritage: "I can't heal people," Lady Strange said at one point, "but I can sometimes take their pain away by laying on my hands."

Lady Strange was a warm, formidable woman of seventy, in the plaid skirt of her Scottish clan, a green velvet jacket (with a small tear stitched up over the left breast), and ruffled cuffs (slightly worn), and she was a stout defender of aristocracy: "It's not the title. What I'm saying is that it's the hereditary influences that will come out. If your ancestors were the sort of people who did the sort of things that got them made peers, the chances are those genes will get carried on."

Kinship was everything. Lady Strange was conscious, for instance, of her connection to Viscount Cranborne, a Cecil and long-time leader of the House of Lords, not just as a fellow Tory, but as "a thirteenth or four-teenth cousin." Such a tenuous connection would surely seem irrelevant to an American, and also to a biologist: The theory of kin selection states that the willingness of one individual to sacrifice for the benefit of another depends on how closely they are related. Or, as a biologist once jokingly summed up the idea, an individual should "gladly die for two brothers, four cousins or eight second cousins." For a fourteenth cousin, on the other hand, most ordinary people would not give a fingernail paring. Yet this abiding sense of family identity was the essence of civilized debate in the House of Lords. Reform, said Strange, amounted to "hereditary cleansing," and as if this genocidal reference had just reminded her, she added, "I can trace my ancestry to Attila the Hun." But the flame faded from her eye as quickly as it had risen. "Not a very pretty ancestor," she admitted. "I think his traits have been rather diluted."

She finished her tea, then queued up at the cash register to pay her bill before returning to the debate. "Fascinating place, isn't it?" she remarked, as if she were only a visitor there herself. Over a shoulder she noticed a full-length portrait of King Henry VII.

"My ancestor," she said.

The Next Worst Thing to Chinese Grammar

By coincidence, that afternoon was also the publication date of a book almost as charming and anachronistic as the House of Lords itself. *Burke's Peerage and Baronetage*, a record of the noble and ignoble ancestors of the

British aristocracy, was making its first appearance in twenty-nine years, having generally been given up for dead and disgraced. The new book was the genuine article, the 106th edition of *Burke's* since its founding in 1826, produced by a team of nine editors at a cost of $1.2 million. "It has more words than the Bible," the publisher boasted, to which his chief editor immediately added, "I would point out that the genealogies are also more reliable than Shem begat Mo." The two-volume set, listing two thousand families and weighing in at twenty pounds, added up to a considerable labor of love.

From a naturalist's perspective, *Burke's* was interesting as a field guide to a rich, successful, and highly cohesive group of animals. But it was also much more than that. For members of the group, *Burke's* was once a reference work for weeding out fakes and keeping up with "noble kinsmen." In its prime, it was also the aristocracy's studbook, for marrying well (meaning one another). In Oscar Wilde's 1893 play *A Woman of No Importance,* Lord Illingworth describes "the *Peerage*" as "the one book a young man about town should know thoroughly." The connection to other members of the group was so important that Lord Marlborough obliged Consuelo Vanderbilt to read up on the two hundred families "whose ramifications, whose patronymics and whose titles I should have to learn." And this was on their honeymoon.

Burke's was the embodiment of the group's extended idea of family. It was the medium by which members signaled worthiness to one another and simultaneously shrouded themselves in an aura of inscrutability and mystification. Outsiders might not know that the peerage consists of five ranks (dukes at the top, followed by marquesses, earls, viscounts, and barons, very much in that order) or how properly to address each, beyond the ritual groveling and tugging of forelocks. Outsiders might also be discouraged (and indeed, were *meant* to be discouraged) by the convoluted nature of aristocratic breeding. Titled families had a bewildering knack for intermarrying with other titled families and banging surnames together with a hyphen or simply switching to new surnames as they inherited new titles, sometimes three or four in a lifetime. The one-name ideal somehow eluded them. The current Duke of Bedford, for instance, is also Marquess of Tavistock, Earl of Bedford, Baron Russell, Baron Russell of Thornhaugh, and Baron Howland of Streatham.

But his real name is John Robert Russell, and his friends call him Ian.

Burke's also helped to hide the members of the class from untutored eyes with its inscrutable language. An entry about the Earldom of Caithness, begins "William, of Mey; dvp unm (strangled by his bro the Master)." The abbreviations stand for "*decessit vita patris*" (died in his father's lifetime), "unmarried," and at the hand of his own brother. This mix of the cryptic and the titillating is typical of *Burke's*, often with some exotic tribal custom tossed in as a bonus: "ROBERT CRICHTON, 8th Lord Crichton of Sanquhar; b c 1568; RC; lost an eye . . . in a bout with a fencing master, John Turner . . . plotted Turner's murder for seven years, effecting it through the agency of one Carlyle. . . . Carlyle and several other accomplices were hanged by a hemp rope but the 8th Lord was granted a peer's privilege of being hanged by a silk one." The inscrutability of *Burke's* was a product of its original function, which was to make the aristocracy seem more exclusive. A British writer, W. H. Mallock, contemplating the relationship between birth and riches among the British gentry, remarked, "Excepting Chinese grammar, I doubt if anything is more complicated."

Powerful Fiction

If *Burke's* was intent on excluding the unwashed, it was in its heyday also happy to publish any wild tale a family cooked up to establish its ancient roots. One historian summarized the resulting book this way: "Impossible men with impossible names . . . do impossible acts in impossible places at impossible times."

Wilde called the *Peerage* "the best thing in fiction the English have ever done." The Feildings (or Feldens), for instance, rose from plain country folk to the Earldom of Denbigh through a good marriage in the 1600s. They subsequently revealed the happy news that they were, in fact, descended, by way of Rheinfelden, Germany, from the imperial House of Hapsburg. Unfortunately for the Feildings, a genealogist named J. Horace Round was making his career in the late nineteenth century exposing what he variously described as "evil work," "gross usurpation," "audacious concoction," and "absurd fiction" in *Burke's Peerage*. Round demonstrated none too gently that the Hapsburg connection was based on "a pack of forgeries," with the result that the Feildings now refer to themselves merely as "Perhapsburgs."

The Churchills, whose twentieth-century descendants were to include Princess Diana as well as Sir Winston Churchill, improved on their ancestry in similar fashion—and also came to grief at the hands of J. H. Round. Not content merely to be descended from the brilliant eighteenth-century general John Churchill, the family felt obliged to drive its roots deeper into history in the person of Sir Bartholomew de Chirchil, who supposedly held the castle of Bristol for King Stephen in the twelfth century and died fighting in his cause. Round gleefully quoted "ancient songs" about Sir Bartholomew written in pseudo-Middle English: "Into the Battail for to fight / He then did make his way; / Ne was there founden any Wight, / So stout as might him stay." Round demonstrated that this noble ancestor was a fiction, pointed out that the castle of Bristol had been held not by King Stephen, but by his enemies, and unkindly concluded that the actual ancestry of the Churchills had been "singularly undistinguished."

Those sorts of tales have long since been purged from *Burke's*. In fact, *Burke's* now reads as if it had been edited by J. H. Round himself. Of Sir Edward George Bulwer-Lytton, for instance, it reports that he was the author of *The Last Days of Pompeii* "and other fictions, not least the account he used to give of his own ancestry, claiming that BULWER was a corruption of 'Bolver', the name the Norse God Odin assumed in battle." Burke's now also lists some of the innumerable illegitimate children of the peerage, and when I talked with Charles Mosley, the current editor of *Burke's*, he guessed conservatively that over ten generations, 40 percent of peerage families had experienced illegitimate descent or what biologists politely termed "discrepant paternity." In other words, the peerage itself was the real fiction.

During the course of my visit with Lord Bath, the conversation turned at one point to a venerable old soul who showed up a few years ago, not in the Thynn family tree but in a cave on the family property. "Cheddar Man," who died 9,000 years ago, is the oldest known complete human skeleton in the British Isles. Researchers have taken DNA samples from the skeleton and from local people to sort out how they might be related. The results from Lord Bath's own cell sample seemed to delight him: He had no discernible connection to Cheddar Man, but "in Her Majesty's prisons," he said, "they take DNA from some very terrible criminals, and mine is identical to one of them." Doubtless another descendant of Edward VII.

Meanwhile, Lord Bath's butler, Cuthbert Barrett, had joined us. He was dressed not in livery but in a comfortable cardigan, and the two of them talked as something like equals. It developed that Barrett had also given a sample for the Cheddar Man study. His DNA turned out to be from one of the most ancient groups known in Europe. It made me think of something an old critic of *Burke's* once wrote: "People who talk about old families sometimes forget the obvious fact that one family is really as old as another. . . . The only difference is that the 'old' family knows, or thinks that it knows, who its forefathers were at a particular time."

Dynastic Tales

But if the peerage was a fiction, it also worked. If blood connections, like Winston Churchill's theoretical one-two-hundred-and-fifty-sixth share of John Churchill's genome, were biologically meaningless, the myth of aristocracy nonetheless endured. It inculcated in its members the sense that they were naturally rich and rulers by right. It enabled them to believe, as Lord Curzon put it shortly before World War I, that "all civilization has been the work of aristocracies." The myth by itself probably wasn't enough to keep the aristocracy in power. But its members also supplemented the myth with the practical step of giving each rising generation "impeccable connections, and an early entry into public life," to go along with the advantages of a good name.

This is a dynastic recipe still closely followed by the families of the rich, though never since on such a scale or with such success. Like the British peerage, families who manage to hold onto wealth and power over generations often do so by cultivating a good story, a family persona— the Rothschilds as bankers and as connoisseurs of great art, fine wines, and fleas; the Rockefellers as philanthropists. The story may change over time and take on a dynamic of its own. The Kennedys, for instance, made their reputation as politicians who grasped life firmly by the buttocks. The Bushes made theirs as members of a New England blue blood establishment reared to regard life from the neck down as an alien empire. "The dynasties have always moved in opposite directions," *New York Times* columnist Maureen Dowd writes. "The Bushes were trying to de-Anglicize and lose the silver spoon while the Kennedys were trying to Anglicize and seize it. The ambitious adventurers wanted to seem like diffident Waspy aristocrats and vice versa." Coincidentally, both families pursued the Rothschild strategy of dispersing offspring to distant capitals to create a dynasty of interconnected centers of power. But the Bushes did it better: Governor Jeb Bush of Florida was thus in position to help hand the 2000 presidential election to Governor George W. Bush of Texas.

A family story provides a sense of collective identity, like the group smell of a house mouse colony. A shared interest, like the Rockefellers' search for better ways to feed the world or fight disease, also gives people a practical motive to come back together periodically and to remain a part of the family. Family is, in a sense, a voluntary relationship. Some children and grandchildren walk away, change their names, do everything in their power to escape the family reputation and live as individuals. (Psychologist Joanie Bronfman talked with one heiress who had herself sterilized to avoid becoming merely a link in the chain. "The real issue," she said, "was that I felt this ancestral demand on my womb.") Others, like the ninth duke of Marlborough, gladly subsume themselves in the family myth. In-laws, like Consuelo Vanderbilt, must likewise choose to resist the myth of their new family, surrender to it, or seek some middle ground before they vanish into the genetic weave.

The strategically more sophisticated families acknowledge the voluntary nature of the relationship. They sustain themselves by supplementing blood connections with a looser, almost tribal sense of identity, a collective loyalty to family legend. "There never was a family of the blood, only families of affinity," says Jay Hughes, a lawyer and trust fund officer, who makes his living telling rich families how to escape the shirtsleeves-to-shirtsleeves trajectory. "What is a family? Two humans choosing by affinity. Every family starts with two names. And the critical component of great families is their ability to perceive that they're families of affinity." After three or four generations, even the most powerful families sort themselves into clans, says Hughes, and "some or all of those clans choose to come back and form tribes," if circumstances suit them.

The Laird Norton family in Seattle is one such tribe. They first became wealthy in the timber industry 150 years ago. "The seventh generation children are on the ground," says Peter Evans, the family president, and the plan is to remain "financially strong and intellectually prosperous" for at least seven more generations into the future. "Having a 150-year goal changes how you act," says Evans. The financial core of the family is a holding company with a portfolio of "legacy businesses," particularly in building materials, and annual revenues of $1.3 billion. Even in Seattle, the Laird Norton Company is largely unknown, "and that's a good thing," says Evans. On the other hand, the family works hard to ensure that its 350 or so members understand their collective identity. It holds an annual family meeting and reimburses travel costs for any family member who owns a share in the company. Affinity is everything: In-laws and even divorced spouses are welcome to own shares, though they're restricted from selling them outside the family. Evans himself is a member of the family only by marriage. During the family meeting, kids attend a kind of camp together, and adults teach one another courses in their areas of interest, from estate planning to sanskrit, at what they refer to as Laird Norton University. Evans welcomes fourteen-year-olds into the family in a formal initiation ceremony, "acknowledging them for who they are, what are their hopes and passions and dreams," and then "we celebrate the richness of the family because we have such a diversity of people and interests." There's also a ceremony "at the other end," acknowledging elders, who stay involved by telling stories and mentoring the young.

"I'd like to know what makes this work," says Evans. "I think it comes down to something simple: We like to be part of something larger than ourselves."

Family Happiness

We like it especially when there's a great deal of money involved. But the Laird Nortons appear to be one of the exceptions, a rich family where the money actually helps bring people together—if only literally, by paying their airfare—and where the family works openly to prevent monetary differences from driving people apart. Otherwise, wealth and a healthy family life seldom sit together at the same dinner table. "As a law of nature, wealth tends to separate everybody," says Peter White of Citibank. He regularly counsels wealthy families to follow the example of the Laird Nortons, by shifting their emphasis away from material things to qualities that are more individual, more personal. But he admits, "I'm not aware of a single situation in which wealth brought people together."

Money, more than the absence of it, seems to turn family members into means to an end. One of the most poignant things about Conseulo Vanderbilt's life, for instance, wasn't simply that her mother Alva forced her to marry a man she did not love, but also that Alva subverted the girl's childhood in this dynastic cause. "A horrible instrument was devised which I had to wear when doing my lessons," Vanderbilt recalled. "It was a steel rod which ran down my spine and was strapped at my waist and over my shoulders—another strap went around my forehead to the rod. I had to hold my book high when reading, and it was almost impossible to write in so uncomfortable a position. However, I probably owe my straight back to those many hours of discomfort."

In such families, children often become little more than "necessary ornaments," and inevitably imperfect ones at that. Because work and social life preoccupy the parents, child care tends to get delegated to outsiders. J. Paul Getty, for example, never mastered the challenge of being a parent and a megalomaniacal empire-builder at the same time. He felt obliged to remain in Europe while his twelve-year-old son underwent three difficult operations for a brain tumor. The boy died during the third operation, but Getty still didn't go home. Like many rich people he found consolation less from family than from the undemanding company of his pets. When Getty's dog Shaun subsequently developed a tumor, Getty flew in the best veterinary surgeon. The dog died anyway, and Getty stayed in his room for three days weeping.

In the families of the rich, lavish gifts often serve as substitutes for love, or as "compensatory enrichment gestures," in one recipient's chillingly clinical phrase. The children grow up baffled, and they struggle to make some more or less generous sense of their family life. Consuelo Vanderbilt recalled that she used to lie in the perfectly appointed bedroom of her mother's house in Newport and reflect "that there was in her love of me something of the creative spirit of an artist—that it was her wish *to produce me as a finished specimen framed in a perfect setting,* and that my person was dedicated to whatever final disposal she had in mind."

The laws of nature—our attraction to kin combined with our attraction to individuals of high-rank—may not, after all, mean that family members love each other any better when they become rich, merely that they watch each other more closely. We want to know who's sitting in the best spot beneath the palm tree, which favored member of the family gets to sit nearby, who's grooming whom and who's getting glowered at, and above all who has the best pile of palm nuts. Little causes for resentment invariably creep in. A man dies in the 1980s and leaves two grandsons relatively equal bequests, an Andy Warhol painting for one and a Roy Lichtenstein for the other. When the two brothers eventually come to sell the paintings years later, Lichtenstein's stock has risen nicely and "Ball of Twine" sells for $4 million. But Warhol's stock has soared, and "Orange Marilyn" goes for the artist's record-high price of $17 million. By all objective standards, both men should rejoice. But it is merely human nature to cast sidelong glances at how the other guy is doing, and more so when he's a sibling.

In another wealthy family, a father remarried, and his new wife was troubled by the financial chasm between them. "If you're not comfortable, how about I give you half my wealth?" the man said, with an heir's insouciance. The result was that, after their deaths, the wife's two children wound up being vastly richer than his own six kids. The relationship between the two sides was permanently tainted with the sense of relative deprivation. Even the most egalitarian gestures contain the seeds of dispute. In the Nordstrom family in the year 2000, for instance, six cousins all held the title of president in the family's struggling department store company. At such times, the cruelty of primogeniture might be kinder for all.

Often, the stakes are high enough, and the accumulated sense of injury so intense, that families end up putting their disagreements nakedly on display in the courtroom: Ronald versus Bruce Winston fighting for the wealth of jeweler Harry Winston, Bill and Charles Koch waging sibling litigation for twenty years for the family oil fortune, the Binghams in bloody tribal warfare over the fate of their Kentucky newspaper fortune; even that most unnatural of rivalries, J. Paul Getty suing his mother, and insurance mogul Saul Steinberg's mom suing him.

Inheritance Power

If the rich have more cause to disagree, maybe, like high-ranking rhesus macaques, they also reconcile more readily. But there is precious little evidence of it. On the contrary, murder, or at least the fear of it, often follows great wealth and power. In 1676, an aristocrat named Mme de Brinvilliers took a chemist named Sainte-Croix as a lover, and he supplied her with a powder to resolve certain otherwise intractable disputes in her family. She tested the remedy first by serving as a volunteer in a charitable hospital and administering it to her patients, who gradually declined and died. When she had worked out the dosage for steady, unobtrusive poisoning, she put the powder to work in her own family. The case would ordinarily be unremarkable; Mme de Brinvilliers was duly caught, tried, and executed for murder. But when Louis XIV subsequently launched a wider investigation into the use of what had come to be known as "inheritance powder," meaning arsenic, no fewer than 442 upper-class men and women were implicated.

A heightened awareness of such peril within their own families still sometimes haunts the rich. For instance, when J. Seward Johnson's many heirs were settling their dispute over the distribution of his wealth, his widow Basia, the former chambermaid, wanted Johnson's children to agree that part of the settlement was to take place only on her death or "upon the expiration of her actuarial life, whichever came later." Lawyers for the opposition referred to this as the "don't-murder-me" plan. Basia Johnson said simply, "I have to protect myself."

Sooner or later, even the very rich must die, though usually in the hope of a dynastic afterlife. In England, a multimillionaire property owner named Nicholas Van Hoogstraten is currently building a monumental house in Sussex for the sole purpose of being buried there (possibly on the theory that this will help posterity forget that he once spent four years in prison for a hand-grenade attack on a delinquent debtor). In the United States, the rich widow Courtney Sale Ross apparently intends to go to her rest in Green River Cemetery in the Hamptons. She has purchased 110 lots there, in pursuit of posthumous territoriality. The next three or four generations of her family will thus be able to moulder together in a little family estate, safely separated from hoi polloi of less refined aesthetic taste.

Death promises eternal rest, but this may be false advertising. Often, the family members gathered around are merely awaiting their chance to scrabble at the bones of the dearly departed. When she died in 1914, Josie Arlington, proprietor of one of New Orleans's most celebrated bordellos, arranged to be buried in a handsome $15,000 pink marble tomb at Metairie Cemetery. (When the founder of the Armour meat fortune had his monument built of pink granite, everybody said it made them think of ham.) But the point of a sumptuous burial isn't simply to leave behind a territory; you must also leave a fortune, a family legend, and a dynasty of relatives willing to defend it. Unfortunately, Arlington's family merely inherited her knack for selling whatever came to hand. They turfed out her sorry carcass and sold the pink marble mausoleum to the highest bidder.

The tobacco heiress Doris Duke, on the other hand, knew exactly the sort of relatives she was leaving behind, and she gave them no such opportunity. After her death, one family member remarked, "I don't care whether [the butler] snuffed the old girl or not. She was a selfish, self-centered old bitch who never did anything for anyone."

Duke prudently opted for burial at sea. She had spent her life among the rich. And now, she told her friends, "I want to be eaten by sharks."

Epilogue

A How-To Guide for Alpha Apes

*Nature, Mr. Allnut, is what we are put in this world to rise
above.*

—ROSE THAYER *(Katharine Hepburn)*, The African Queen

ONE GOLDEN AUTUMN MORNING IN LOS ANGELES, I WENT CRUISING
through the hills in a bright red Ferrari F355 Spider convertible. The guy
who rented it to me gave me something between a pep talk and a ratio-
nale for paying $1,200 a day to rent a car: "You want people to look at
you. You want people to say, 'There goes somebody important.' " The
first thing I did was try to roll down the window to put my elbow out, but
the sporty curvature of the driver's door meant the window would not go
down all the way. The second thing I noticed was that every time I turned
into a driveway with an apron steeper than 7.5 degrees, the overhanging
front end of the Ferrari hit the pavement with a horrible scraping that
made me tremble for my $10,000 deductible. Ah, but the Ferrari sizzled
on the open road. It went winding up into the hills like a snake on the
scent of a terrified rat. I pulled up briskly to a stop sign and tried to look
rich or, rather, *comfortable*, as the heat and the fumes from the big engine

in back washed down over me and made me gag. "Hey, man. Great car," someone yelled.

A former Ferrari owner had warned me it might be like this: "It's great for everybody except the guy behind the wheel," he'd said. Driving around that day, visiting expensive homes behind iron gates, a lot of things about being rich seemed the same way: Nicer to look at than to live with. The cars, the homes, the art, the fashionable friends were trappings put on for show, often at the expense of comfort, and what they inevitably begot was not satisfaction, but a creeping need for better trappings. The rich people I met seemed to rattle around their vast enclosed spaces, like zoo animals in their cages, yearning for something more, something other—not the food dished up for their pleasure, but the hunger and the chase.

At one grand home, I waited in a sitting room and passed the time wandering around looking at the clutter of handsomely framed photographs of my hostess arm-in-arm with some of the great boldface names of her day. I counted forty-five pictures of herself on display in the room, not including photo albums. Then she came grandly in and started talking about herself with such warmth that, when I finally got away two hours later, I realized I had never gotten to tell her why I was there in the first place. The rich were some of the neediest people I had ever met. It made me feel bad about likening them to chimpanzees, and not just for how poorly it reflected on the chimps.

The whole idea of camouflaging myself as one of the rich seemed absurd. The Ferrari registered dimly, if at all, on their consciousness. It was the sort of car they gave the kid for high school graduation. It was also clearly the wrong car for me. The badge of status didn't fit. I felt like one of those barn swallows on whom biologists have glued an unduly long tail. Or maybe I had a blinking neon sign on my forehead, and what it said was not NOUVEAU RICHE, but RENTAL.

So I generally ended up telling them who I was and what I was up to (at least when I had the chance to get a word in edgewise). "You mean you're looking at us as if we were monkeys?" a woman asked me one day, across the table at a benefit lunch featuring a fashion show by Escada. "Well, apes," I replied, as if this clarification would make it any better. She smiled with evident delight, and then everyone joined in, seeking details.

I think they had been watching too much Discovery Channel. I was afraid someone was going to launch into a disquisition on sperm competition, along the somewhat reductive lines of the actress Ashley Judd, who recently appeared on a late-night talk show and summed up evolutionary psychology in a phrase: "Women want as much sperm as possible."

They were, in any case, comfortable with the idea of themselves as animals, though they suffered from a lingering suspicion that a natural history point of view might prove less than edifying. (St. Bonaventura clearly implied as much when he remarked, "The higher a monkey climbs, the more you see of its behind.") What they wanted to know was: What good could a natural history of the rich do for them? What could it teach them about better ways of being on top? What ten leadership lessons could they take home from Tamba the Ape or even perhaps from Harry the Hangingfly? Better yet, could I give them the executive summary on that?

So let's start with the good news: What biologists have to tell us about the cultural subspecies *Homo sapiens pecuniosus* is at least more edifying than Sigmund Freud's ideas about wealth. Freud equated money with excrement, and his followers have frequently characterized wealth accumulators as "anal types" driven to pile up cash and commodities by their stunted emotional development and unresolved conflicts over toilet training. "Cupidity and collecting mania, as well as prodigality, have their correlating determinants in the infantile attitude toward feces," one such disciple has written.

Freud's scatological point of view certainly makes sense, for dung beetles: Little Stephanie Dungbeetle likes to burrow under a fresh cow pat to lay her eggs, and she uses cow dung as food for her young. Her mate, with a formidable horn on his snout, stands guard outside the entrance to the family pile. (Meanwhile, underdog males, born without the dung to grow proper horns, are reduced to sneaking in from the side for little trysts.) But the equation of money with excrement is dubious at best for humans. Anyone who has spent much time with small children will recall that they are by and large indifferent to their own excrement, sometimes dismayingly so. The primal hunger for food, on the other hand, is loud, insistent, and unrelenting. Given that the business of science is to prefer a simple explanation over an unreasonably convoluted one, it makes better sense that we value money as a symbol not of excrement but of bread.

The Freudian argument that collecting and accumulating represent substitutes for sexual conquest is somewhat more plausible. "Every collector is a substitute for a Don Juan Tenerio," Freud wrote, "and so too is the mountaineer, the sportsman, and such people. These are erotic equivalents." Surely it would have been more persuasive to argue that sexual sublimation is the driving force merely for *some* collectors, notably Freud himself. The Comtesse Anna de Noailles, on meeting Freud and getting a first-person sense of his bottled-up sexuality, remarked: "Surely *he* never wrote his 'sexy' books. What a terrible man! I am sure he has never been unfaithful to his wife. It's quite abnormal and scandalous." For the rich, adventures in art or mountain-climbing often intertwine effectively with a penchant for sexual indulgence; it's about synergy, as they say. The idea that collectors and sportsmen are merely sublimating their sex drives seems positively quaint in the age of the tell-all autobiography. It is far simpler to explain the urge to achievement in these fields by analogy to the peacock's fan and other forms of sexual display behavior in the animal world.

Suggesting that biologists are more enlightening than Freudians about the nature of wealth may seem like damning with faint praise. Yet the biological perspective also offers useful clues both about how the rich should live and about how the rest of us should live in a world run by rich people. If we put on our self-help hat, we can in fact arrive at something like *An Alpha Ape's Ten Rules for Living Wisely in an Imperfect World.*

1. **Get to know the Three Big Lies of the subspecies.** You will want to use them at regular intervals: "I'm not really interested in money"; "Power doesn't matter to me"; and "I don't give a damn about impressing other people." The truth is of course the opposite: Control of resources, social dominance, and effective display behavior are what it's all about, for bare-assed monkeys and Brioni-clad humans alike. Good manners (and the need to disarm potential critics) merely oblige us to pretend otherwise.

2. **Make friends shrewdly.** Whether one already has wealth and power, is merely seeking it, or is squirming under its hairy thumb, the

essential rule for social primates is to cultivate useful allies. Send your children to the right schools, live in the right neighborhoods, give generously to socially desirable charities. The houses on Red Mountain in Aspen are frequently owned by people living far beyond their means, and the plan is that they may soon be able to afford those homes if they can just hang around long enough with the right neighbors at the right parties. This plan makes sense, at least if they time their climb up the social ladder so that their financial burn rate doesn't incinerate the rung beneath their feet. Meeting rich people on the golf course or at a party is infinitely more effective than trying to waylay them at the office, because it helps position you as a member of the subspecies. On the other hand, it's important not to be too obvious about why you're there (see Rule 1). Don't, for instance, flatter your wealthy neighbor to his face, which might raise his mistrust. Flatter him to his best friend and let the word work its way back. (The corollary of this rule is: *Watch out for your neighbors*. Red Mountain has been home to some of the most spectacular swindles of our day.)

3. PAY ATTENTION TO WOMEN. Their power may be more effective than that of their male counterparts' for being less obvious, and they live longer.

4. GIVE EARLY. GIVE OFTEN. GIVE MORE THAN YOU CAN AFFORD. Social primates are ingenious at keeping score, and someone is always noticing. The friends you make and the goodwill you accumulate may save your life. Los Angeles marketing entrepreneur Robert H. Lorsch regularly gave large sums to the John Wayne Cancer Institute, even when he was living on credit cards. At fifty, he went for a thallium treadmill test and his regular doctor told him his heart was fine. But his gifts had gained him the friendship of the Cancer Institute's chief researcher, who offered to take a look at the results; he noticed that Lorsch was suffering from an aggressive medullary thyroid cancer. Without charity, Lorsch figures, he might now be dead. (On the other hand, once someone has saved your life, this also goes on the scorecard. So now Lorsch may need to triple his giving.)

And of course, *always* give to politicians, particularly when they are on the way up to alpha. *The New York Times Magazine* recently reported the case of Missouri coal investor Irl Engelhardt and his associates at Peabody Energy, who gave $700,000 to help elect President George W. Bush and Vice President Dick Cheney. The procoal energy plan duly produced by the new Administration helped make Peabody's initial public offering a huge success. Engelhardt's take alone was worth more than $23 million.

5. **PUT ON YOUR PLUS FACE.** In all primates, a confident posture is a self-fulfilling prophecy of success. Throw back your shoulders and walk straight, and people will get out of your way. It's also possible to condition yourself for success, by starting with small victories and working upward: Bone up on market-neutral arbitrage funds and make a telling point over cocktails. Find a tennis partner you can beat and thrash him. (Choose someone you're not quite sure you can beat; the body is not fooled by cheap victories.) Get physical. Think of Chinese fighting crickets. Researchers at the University of Leipzig in Germany studied the ancient custom among fight aficionados of shaking a defeated cricket in their clasped hands and tossing it into the air several times before putting it back into the ring to fight again. They found that triggering the flight muscles somehow "resets" the cricket's aggressiveness. Some similar factor may be one reason rich guys are always looking for pickup basketball games or other forms of physical competition; even dour John D. Rockefeller liked to drag race with his team of trotters. Start by boxing with your own shadow and in time you may be ready to go straight to the gut in a financial smackdown with former General Electric boss Jack Welch.

6. **STRIKE DECISIVELY AND WITH OVERWHELMING FORCE.** When you have lined up your allies and seem certain of their support, move ruthlessly and without warning against your superiors. Anything less risks being like the anti–grizzly bear spray sometimes sold at national parks, which comes, as a researcher friend once put it, only in "the piss-em-off size."

7. **WIELD YOUR NEW STATUS GENTLY.** This is the single most important thing a natural history perspective should teach us: In all things primate, dominants prolong their elevated status—and may also enjoy it

more—when they look out for the best interests of their subordinates. Humans seem to be "ethologically despotic," like chimpanzees; that is, we have a natural predisposition to hammering other people into submission. But primatologist Christopher Boehm, who has studied subordinate behavior in chimpanzees and in human groups, argues that over the past one hundred thousand or so years we have become culturally egalitarian. His argument has a wide streak of wishfulness to it: "I believe that if a stable egalitarian hierarchy is to be achieved," he writes, "the basic flow of power in society must be reversed definitively." This unfortunately makes it sound like some kind of left-wing plot. The truth is that laws, marital practices, labor unions, scrutiny by the media, and other factors have already gone a considerable distance toward reversing the flow of power, and smart despots learn to adapt.

Even Louis XIV, whose reputation for absolute rule is incarnated in the apocryphal phrase "*L'état, c'est moi,*" paid considerable attention to the well-being and goodwill of his people. He once wrote: "We must consider our subjects' good before our own. They are indeed like a part of ourselves since we are the head of the body, and they are the limbs. We must give them laws for their own advantage only; and we must use the power we have over them only so as more effectively to bring them happiness." Looking out for the interests of subordinates does not mean giving them everything they want, nor does it preclude ruthless or self-serving acts, but benevolent despots find ways to distance themselves from their bloodier deeds. If you are frat-boy friendly, you hire a mad dog to do your dirty work. If you are a mad dog, you need a frat boy or sorority sister to staunch bloodshed and soothe the injured.

8. FAMILY SHOULD ALWAYS COME FIRST. When you have money, everybody wants to be your baby. This is your nightmare and of course your temptation. Your subordinates treat you like Big Daddy and do exactly what you want; they line up to lip the burrs from your hindquarters. Spouses and children, on the other hand, loudly note your flaws and balk at your bidding. What to do? Even baboons are smart enough to stick with their real family. Peter White of Citibank also counsels the rich to keep company "with people who've known you for years and know what a doofus you really are."

As to the shirtsleeves-to-shirtsleeves trajectory and the future of your dynasty, this is your big call in the grand genetic crapshoot. The world of the rich is half-filled with experts explaining how to preserve your wealth for future generations and half-filled with charitable fund-raisers explaining why the smartest thing you could do is give it all away. The argument for giving it away is that your genetic future may be more secure if your kids know from the start that they have to make it on their own. Give them your love and your good name, kibitz on their behalf like Grandma Vervet, and leave them nothing else; they may turn out to be scrappers like you. But kin selection is a powerful force. So most families opt to preserve capital and build dynasties.

The kids may still turn out sane. All you have to do is spend time with them. Also, make them work for every dime when they are young, keep them on a budget, limit extravagance, and don't start to turn over the family fortune until they have already established themselves in their own careers.

9. BUILD A MODEST HOUSE. Did anyone really envy William Randolph Hearst in his 110 rooms at San Simeon? As a practical matter of being comfortable and a little grand at the same time, a 5,000-square-foot house is just about too much. The children should not need a map to get to their parents' bedroom. If you must build a monster house "for the art collection," then build a guest house out back and live there. This strategy for simultaneous show and comfort is surprisingly common among the rich. *The New York Times* and *Forbes* have both lately featured articles on very upscale treehouses, including one where a Microsoft couple go to practice qi gong and the occasional pig grunt.

10. PLAN YOUR ESCAPE ROUTE. Never walk in a door or agree to a deal unless you know two or three ways back out again. Animals are very nervous about being cornered, and you should be, too. This is undoubtedly one reason yachts and private jets are so popular among the rich. Throughout his life, for instance, J. Paul Getty steadily traded upward for larger and larger yachts, but not apparently for love of the sea. According to biographer Robert Lenzner, he "was worried about the Communists" and told his wife Ann Rork that "he needed to know that he could sail away at any time if he was forced to escape from the enemy who might want to take his millions away from him."

His fear of Communists was misplaced. The more immediate threat to the rich is of course other rich people, who are constantly jockeying like beta males for the chance to displace their alphas. This is why one house in Incline Village, Nevada, has "his-and-her escape passages"— because of a threat from a former business partner. And perhaps also to let husband and wife escape from each other. Both spouses in this happy marriage pack handguns.

One time on the Texas border, I visited the home of a lawyer. He had a stagger-wing biplane parked in one corner of his living room, a Stearman trainer in another, and a Cessna 180 in the middle. We climbed into the Cessna, and one wall of the house rolled up. Then we taxied out past the livestock on the lawn (peacocks, appropriately), onto the runway in back, and took off. We passed over a mansion backing up onto the Rio Grande, home of a local surgeon and trophy hunter. "He's got more animals on the walls of that house than they have in the San Diego zoo, and that's no lie," the lawyer said. Then he pointed a mile or two ahead, to a squatter village in the battered desert hills across the river, where newcomers up from the Mexican interior were building their homes with cardboard and factory pallets and dreaming of someday making dynasties of their own. The houses looked like a scattering of rotten teeth in a damaged old mouth. "Shows you what a fragile society we live in," the lawyer said.

The plane banked eastward. Away from the green swath of the river valley, the bleached-out canvas of desert stretched endlessly to both horizons. We headed for the lawyer's hot springs ranch, his refuge from humanity, and we landed on a rough dirt runway atop a mesa. The only rules of the species that seemed to matter out there were the personal ones posted on a wall at the ranch: "Do not talk politics. Do not talk business. Do not hustle elected officials. Eat when you are hungry. Piss anywhere."

By now, though, I knew better. The rich like to pretend that nature is something they have risen above. But in their hearts, they know rising is a myth. Looking at that sign, I understood that I was merely a guest in the territory of a dominant animal.

What it really said was, "Scent-mark here, and you will be eaten alive."

Notes

Introduction: Naturally Rich

p. 11 Perelman's mate-guarding: Shah 1996.

p. 12 *Johannseniella nitida*: Buss 1994, p. 124.

p. 12 Rockefeller as lion: Chernow 1998, pp. 137–38.

p. 12 Dedman as mosquito: *Forbes,* October 10, 2000.

p. 12 Masayoshi Son: *Forbes,* July 3, 2000.

p. 13 Sunflower quote: De Waal 1982, p. 174.

p. 14 Martha and the nematodes: "Genome Sequence of the Nematode *C. elegans*: A Platform for Investigating Biology." *Science* 282, no. 5396: 2012–18.

p. 14 Asian elephant pheromone: Rasmussen 1996.

p. 15 Bluebird infidelity: Gowaty and Bridges 1991.

p. 15 Jennie Jerome's lovers, estimated by George Moore: Fowler 1989, p. 193.

p. 15 John Strange Spencer Churchill: Manchester 1983, p. 136.

p. 15 "A law of nature": Collier and Horowitz 1976, p. 91.

p. 15 Standard Oil and Darwinism: Chernow 1998, p. 154.

p. 17 The tribal Binghams: Bingham 1989, p. 5.

p. 18 Elephant seal weight loss: Andersson 1994, p. 239.

p. 18 Gutfreund invitations: Winokur 1996, p. 23.

Chapter One: Scratching with the Big Dogs

p. 23 Bezos quote: *Town & Country,* June 2000, p. 157.

p. 26 Vanderbilt mantle: Goldsmith 1980, p. 83.

p. 28 Respectable poverty: Jaher 1980, p. 197, from *The New York Daily Tribune,* March 25, 1888.

p. 28 Median income and $200 fee for the millionaires next door: Stanley and Danko 1996, pp. 9, 192.

p. 28 "Pentamillionaires": *Barron's,* September 18, 2000.

p. 29 The meaning of "millionaire": *The Wall Street Journal,* March 16, 2001.

p. 29 Americans with a $1 million income: *The New York Times,* February 7, 2002.

p. 29 Relative deprivation and social isolation: Barkow 1989, p. 196.

p. 30 Wealthy or comfortable: *The New York Times,* October 15, 2000.

p. 30 Vanderbilt sody water: Vanderbilt 1989, p. 50.

p. 30 Vanderbilt estimated wealth: Goldsmith 1980, p. 106.

p. 30 Peltz background: Bruck 1988, pp. 105–19.

p. 31 Peltz's lament: *The New York Times,* March 3, 2000.

p. 31 Rockefeller on Morgan: Strouse 1999, p. 15.

p. 32 "Public screwing": Griffin and Masters 1996.

p. 33 "So goddam rich": Goldsmith 1980, p. 261.

p. 34 "Cultural pseudo-speciation": Lorenz 1966, pp. 76–77.

p. 35 Peacock pickiness: Petrie, M.; T. Halliday; and C. Sanders. 1991. "Peahens Prefer Peacocks with Elaborate Trains." *Animal Behaviour* 41: 323–31. See discussion in Zahavi and Zahavi 1997, p. 33.

p. 36 "Not, strictly speaking, cannibalism": Lorenz 1966, p. 79.

p. 36 Forbidden genealogy: Shoumatoff 1985, pp. 68–9.

p. 36 Duke of Somerset: Cowles 1983, pp. 404–5.

p. 36 Maharani of Baroda: Allen 1984, p. 206.

p. 37 The Voltaic shock: Thorndike 1976, p. 168.

p. 37 Encapsulation: Barkow 1989, p. 196.

p. 37 The rich man as a god: Stengel 2000, p. 50. See also a similar quote by Cettie Rockefeller to her son in Chernow 1998, p. 357.

p. 37 Soros on wealth as a disease: *Worth,* June 1999.

Chapter Two: The Long Social Climb

p. 39 "Descended from apes!" has been attributed variously to the wife of the canon of Worcester Cathedral (in Ashley Montagu's *Man's Most Dangerous Myth: The Fallacy of Race,* p. 63), to the wife of Bishop Wilberforce by George Washington University biologist David Atkins on his Web site, and to "a joke of the Darwinian period, perhaps from *Punch* magazine" (in *Narrow Roads of Gene Land: The Collected Papers of W. D. Hamilton,* p. 14).

p. 39 From proto-primates to Julia Roberts: Miller 2000, p. 225.

p. 40 Our nearness to our evolutionary roots in Dawkins 1994, p. 84.

p. 41 Six million years: Wrangham and Peterson 1996, p. 42.

p. 41 The inspecting general, and dancing with Wales: Dawkins 1994, pp. 84–5.

p. 41 Stone tools 2.5 million years ago: Diamond 1992, p. 34.

p. 41 Timetable for language, et al.: Wrangham and Peterson 1996, p. 61.

p. 42 Neolithic crops: Lev-Yadun et al. 2000.

p. 44 "Call me back when the parrot's gone": Bruck 1988, p. 112.

p. 44 Rothschilds "like drunkards": Ferguson 1998a, pp. 102–3.

p. 44 "Sexual incapacity": Stove 1995, p. 43.

p. 45 Car ownership and longevity: Goldblatt 1990.

p. 46 Studies on mortality and wealth: Carroll, Smith, and Bennett 1996.

p. 47 Rockefeller's hospital rooms: Chernow 1998, p. 477.

p. 47 J. Seward Johnson's constipation: Goldsmith 1987, p. 6.

p. 47 Holding out to beat estate tax: *Forbes,* April 16, 2001, p. 30.

p. 47 Longevity in dominant animals: Ellis 1995, pp. 268-69, 272.

p. 47 Rockefeller's private hospital rooms: Chernow 1998, p. 478.

p. 48 Bonobos and chimps: De Waal 1997, pp. 4, 6, 24.

p. 48 DNA comparisons: Diamond 1992, p. 23.

p. 49 Sex in chimps and bonobos: De Waal 1997, p. 32.

p. 49 Male-dominated societies: Wrangham and Peterson 1996, p. 118.

p. 50 Lip-smack and backstab: Cheney and Seyfarth 1990, p. 184.

p. 51 Average English vocabulary: Miller 2000, p. 369.

p. 52 Machiavellian Intelligence hypothesis: Dunbar 1996, p. 60.

p. 55 Calvin Klein on Linda Wachner: *Fortune,* September 4, 2000, p. 225.

Chapter Three: Party Time

p. 59 Kozeny background: *Fortune,* March 6, 2000.

p. 59 Kozeny's swimming pool: *Fortune,* December 23, 1996.

p. 60 Kwakwaka'wakh Indians: Barkow 1989, p. 195.

p. 61 The nine-foot-long yam: Dugatkin 2000, pp. 109–10.

p. 62 Parties as bribes: Ferguson 1998a, pp. 200–1.

p. 62 Old Guard of hunter-gatherers: Bender 1978, pp. 206, 211.

p. 63 Food sharing by ants: It may well turn out that even ants are not true Communists. Deby Cassill, a biologist at the University of South Florida, St. Petersburg, recently presented a paper arguing that ant scouts use advertising and salesmanship to recruit comrades. She has also found enormous variation in how much food individual ants seem to get.

p. 64 "A raging sociopath": Gorman, M. L., and R. D. Stone. 1990. *The Natural History of the Mole.* Ithaca, NY: Cornell University Press.

p. 64 Squirrel hoarding: Vander Wall 1990, p. 19.

p. 64 Chimpanzee hunting, food sharing: De Waal 1996, p. 140–41.

p. 64 Individual hunter-gathers amass no surplus: De Waal 1996, p. 137.

p. 65 Ntologi: De Waal 1996, p. 143.

p. 65 Rothschild abstinence: Ferguson 1998a, p. 196.

p. 65 The perks of sharing the kill: De Waal 1996, p. 137.

p. 67 "Triple-A" aggrandizers: Hayden 2001.

p. 67 Staples to status food: Hayden 1990.

p. 68 Selfish elites, ruining the lives of others: Hayden 2001, p. 247.

p. 69 First car race in America: Goldsmith 1980, pp. 87–88.

p. 69 Gaining prestige and social control: Hayden 1998, p. 33.

p. 69 Hayden quote: Hayden 1998, p. 33.

p. 70 Kozeny's yacht: *Fortune,* March 6, 2000, p. 8.

Chapter Four: Who's in Charge Here?

p. 72 Singh quote: Allen 1984, p. 95.

p. 72 An odd little fish: Hoffmann et al. 1999, pp. 14171–76.

p. 73 King Abdullah of Jordan: *Washington Post,* August 11, 1999; *The New York Times Magazine,* February 6, 2000.

p. 74 Thorleif Schjelderup-Ebbe: Price 1995.

p. 74 "Alpha male": *The New York Times Magazine,* November 21, 1999.

p. 74 Larry Ellison quoting Genghis Khan: Wilson 1997, p. 88. The attribution is curious as Genghis Khan left no written records. Perhaps Ellison is channelling his spirit, or possibly he did not like the effect of attributing the remark to Gore Vidal, who has said much the same thing.

p. 75 Dominance and its predictable effects: Bernstein 1981, p. 428.

p. 76 Dominance as a taboo: De Waal 1982, p. 193.

p. 76 Genghis Khan recanted: Smithsonian Institution Oral History, http://americanhistory.si.edu/csr/comphist/le2.html.

p. 76 "Latency to emission": Bernstein 1981, p. 420.

p. 76 Power hum: Gregory and Webster 1996.

p. 77 "An elaborate code": Sapir 1949.

p. 78 Ellison wakes Allen: *The Washington Post,* October 30, 2000.

p. 78 King Abdullah dressing down: *The New York Times Magazine,* February 6, 2000.

p. 80 Rockefeller's megalomania: Chernow 1998, p. 132–33.

p. 80 Alpha personality as "deviant": Boehm 1999, p. 194.

p. 80 Ted Turner's father: Bibb 1993, pp. 14–15.

p. 80 "The love of my family": Branson 1998, p. 16.

p. 80 The Trump family chant: Hurt 1993, p. 13.

p. 81 Intense motivation to dominate: Goodall 1986, p. 425.

p. 81 Mike the chimp: Goodall 1986, p. 428–29.

p. 81 No rules for Onassis: Winokur 1996, p.28.

p. 82 "The game of chicken": Lewis 2000, p. 188.

p. 82 Dominance and size: Dunbar 1996, p. 145; and Etcoff 1999, pp. 172–76. Buss, p. 39; Tahiti chiefs as "different race" in Wason 1994, p. 73.

p. 82 Berlusconi on tiptoe: *The New York Times Magazine,* April 14, 2001, p. 42.

p. 83 Henry Nicholas: *The New York Times,* June 26, 2000; *The Times* (London), March 23, 2001; *Worth,* June 2001.

p. 83 The winner effect: Mazur and Booth 1998, p. 362.

p. 83 Testosterone and success in grade school: Schaal et al. 1996.

p. 85 "Deep croak": Dunbar 1996, p. 144.

p. 85 Orangutan noise: Wrangham and Peterson, pp. 134–35.

p. 85 Dhoom-dham: Allen 1984, pp. 212–13.

p. 85 Mike the chimp's clanging cans: Goodall 1986, p. 426.

p. 86 The plus face: Zivin 1977.

p. 86 Walking tall: Weisfeld and Beresford 1982.

p. 86 Never had a door shut in her face: Bronfman 1987, p. 57.

p. 87 Oracle's insufficient swagger: *Fortune,* November 13, 2000.

p. 87 Dominance as "an invention": Altmann in Bernstein 1981, p. 431.

p. 87 Eye contact: Weisfeld and Beresford 1982, pp. 116–17.

p. 87 The absence of fretting and the intensity of the stare: Zahavi and Zahavi 1997, p. 55.

p. 88 J. P. Morgan's eyes: Strouse 1999, pp. 650–51; and Steichen 1963.

p. 89 Sun moons Microsoft: *The New York Times,* February 28, 2000.

p. 89 Behind every fortune: Balzac 1835, 1954, p. 132; see full text at http://digital.library.upenn.edu/webbin/gutbook/lookup?num=1237.

p. 89 "The honorable predatory impulse": Veblen 1899, p. 141.

p. 92 "The Tiger of Mysore": Forrest 1970, pp. 44–45, 214–16, 290–1; and Ward 1983, p. 82.

p. 93 Squirrel monkey penis display: Morris 1969, p. 108.

p. 94 Trump World Tower: *The New York Post,* June 20, 2001.

p. 94 Clark's yacht: Lewis 2000, p. 26.

p. 94 Katzenburg's Dick: Masters 2000, p. 241.

p. 94 The Big Swinging Dick: Lewis 1989, p. 46.

p. 94 Phallic display in humans and gorillas: Miller 2000, pp. 230–4; and Etcoff 1999, pp. 180–84.

p. 95 Walking upright: Maxine Sheets-Johnstone cited in Miller 2000, pp. 233–4.

p. 95 Vervet monkeys: Eibl-Eibesfeldt 1972.

p. 96 Phallocarps: Diamond 1992.

p. 96 "A ritualized threat to mount": Etcoff 1999, p. 182.

p. 97 Urination syndrome: Aldrich 1988, p. 192.

p. 97 Duke of Marlborough: Fowler 1989, p. 236.

p. 98 Squirrel monkeys in captivity: Morris was interpreting the work of Ploog, in Altmann 1967.

Chapter Five: Take This Gift, Dammit!

p. 100 Gates and smart dog: Manes and Andrews, p. 17; Turner on wolves: *The New Yorker,* April 23, 2001.

p. 100 Arabian babblers: Zahavi and Zahavi 1997.

p. 101 Altruistic amoeba: Strassman et al. 2000.

p. 102 *Forbes* 400 as Super Bowl for the rich: *The New York Times,* August 25, 1996.

p. 102 "Horrible jumble": Balsan 1973, p. 68.

p. 102 Airborne muttonchops: Fowler 1989, p. 143.

p. 103 Wannabe Old Money: This Morgan Stanley Dean Witter ad ran in *The New Yorker* in March 2000.

p. 103 Vervet inheritance: Cheney and Seyfarth 1991, pp. 29–33.

p. 103 Can't shoot the kids: Reuters, September 11, 1998.

p. 103 Turner's UN gift: *The New York Times,* September 19, 1997.

p. 104 "The more money has come in": Turner interview with Larry King quoted by CNN Interactive, September 19, 1997.

p. 104 Getty, Guggenheim, et al.: Thorndike 1976, pp. 88, 101, 179.

p. 105 Stanford chicanery: Holbrook 1953, p. 121.

p. 105 Charles Lamb in *Table Talk.*

p. 106 Sexual adoration: Miller 2000, p. 326.

p. 106 King Victor Emanuel's toenail: Tabori 1961, p. 50.

p. 106 Branson fighting: *Sports Illustrated,* February 15 1999.

p. 106 Weatherhead gushing: *The New York Times Magazine,* June 24, 2001, p. 40.

p. 107 Rothschild as beggar: Thorndike 1976, p. 168.

p. 108 "The Model Millionaire": Ferguson 1998b, pp. 233–34.

p. 108 "Give a beggar a guinea": Morton 1962, p. 64.

p. 109 Hawley on how dominance evolves: Hawley 1999.

p. 111 "The Beast Is Back": *Fortune,* May 30, 2001.

p. 111 Gates v. Allen: Rivlin 1999, p. 63–64.

p. 112 Turner and *Citizen Kane*: Goldberg and Goldberg 1995, p. 186.

p. 112 Murdoch's idea of free speech: Auletta 1997, p. 289.

p. 112 Murdoch getting even: *The Wall Street Journal,* September 27, 1996.

p. 113 Turner's United Nations gift and the gibe at Gates: *The New York Times,* September 19, 1997.

p. 114 Gates spoke to Barbara Walters on ABC's *20/20* on January 30, 1998; his $1 billion gift was reported in the *Seattle Times* for September 16, 1998.

p. 114 Robert Wright on Bill Gates's potlatch: *Slate,* March 27, 2000.

p. 114 The potlatch chant: Eibl-Eibesfeldt 1972, p. 207.

p. 115 Sean Combs's proposed gift: *The Village Voice,* December 12, 2000.

p. 115 Rupert feels sorry for Ted: CNBC interview, May 14, 2001.

p. 116 "Ted Man Walking": *New York Post,* April 5, 2001.

p. 116 The relative net worth of Turner and Murdoch in mid-2001: *Forbes,* July 9, 2001.

p. 116 Murdoch as "scum": David Plotz, *Slate,* May 24, 1997.

p. 116 Comparison to Hearst: Reuters, May 19, 2001.

p. 116 "Look at AOL Time Warner": CNBC interview, May 14, 2001.

Chapter Six: The Service Heart

p. 118 The aerial cow: Ward 1983, p. 54.

p. 119 Good and bad hustlers: Kessler 1999, p. 14.

p. 119 Courtney's aesthetic achievements: *W,* September 2000.

p. 120 Language as a substitute for grooming: Dunbar 1996, p. 78.

p. 120 King Louis XI's fleas: Busvine 1976, p. 71.

p. 120 "Mutual mauling": Dunbar 1996, p. 78.

p. 120 Months or years of sucking up: Cheney and Seyfarth 1990, p. 71.

p. 121 Marcos: Cheney and Seyfarth 1990, p. 41.

p. 121 Louis XIV's mistresses: Cowles 1983, p. 58.

p. 121 Dennis Tito: *USA Today,* May 1, 2001, p. 1.

p. 121 "Mercy guard me!": Milton's *Comus*, lines 695–705.

p. 121 Castlehaven scandal: Sykes 1982, pp. 61–66.

p. 122 Eleven-foot tigers: Allen 1984, p. 141.

p. 126 Our attraction to dominants: Eibl-Eibesfeldt 1972, p. 120.

p. 126 J. Seward Johnson as a sexual abuser: Goldsmith 1987, p. 18.

p. 127 Bonding through a common enemy: Eibl-Eibesfeldt 1972, p. 164.

p. 127 The Chimp Channel: *Toronto Sun,* June 9, 1999.

p. 127 Our desire for order: Eibl-Eibesfeldt 1972, p. 168.

p. 127 Milken as "Dad": Griffin and Masters 1996, p. 143.

p. 127 Undisturbed pecking orders lay more eggs: Bernstein 1981, p. 433.

p. 128 A different kind of animal: Barkow 1989, p. 196.

p. 128 Submissive displays: De Waal 1982, p. 87.

p. 129 Lady Montdore genuflects: Mitford 1949, p. 92.

p. 129 "Wealth doesn't change people": *The New York Times,* June 26, 2000.

p. 129 Rank and serotonin: Masters and McGuire 1994, p. 130–45.

p. 129 Vervets and one-way mirrors: Raleigh and McGuire 1984, p. 408.

p. 130 Jean-Marie Messier: *Financial Times,* July 29, 2000.

p. 130 Henry Fok as "The Leader": *Worth,* June 1996.

p. 131 Smelling rich: Andersen 1999, p. 47.

p. 131 Cortisol spiking, subclasses of subordinates: Virgin and Sapolsky 1997.

p. 132 "I'm a dildo, Harvey": *Vanity Fair,* July 2000.

p. 132 Cortisol, socioeconomic status, and the tubby gut: Adler and Epel 2000.

p. 132 Rank and reproduction in gelada baboons: Dunbar 1996, p. 41.

p. 133 Wet-nurses sacrificing their own children: Trexler 1973. See also Dickemann 1979, p. 353; and Wrangham and Peterson 1996, p. 235.

p. 133 The death of a "fine fat pink baby": Gathorne-Hardy 1973, pp. 39–40.

p. 133 Chinese eunuchs: Gulik 1961, pp. 255–56.

p. 134 Sean Combs's assistant: *Vanity Fair,* July 2000.

p. 134 Hereditary pruning among the Binghams: Bingham 1989, pp. 214, 245.

p. 136 Hiring Civil War substitutes: Chernow 1998, p. 69.

p. 136 The "twenty nigger" law: Mitchell 1988, p. 160.

p. 136 Rank determining who lives, from Tikopia to *The Titanic*: Boone and Kessler 1999.

p. 137 Aboard the *Titanic,* Guggenheim in evening clothes: Birmingham 1967, p. 274; Isadora Straus, Lord 1955, p. 79.

p. 137 First-class heroism: *The Bulletin,* San Francisco, April 19, 1912.

p. 138 Kerry Packer's transplant: *The Scotsman,* November 24, 2000; and *The Daily Mail,* November 23, 2000.

p. 139 "Laugh after he laughs": Stengel 2000, pp. 51–2.

p. 139 Mike the Chimp's imitator: Goodall 1986, p. 426.

p. 139 Yerkes experiment: Lorenz 1966, pp. 42–43.

p. 140 Imitating injured alpha: De Waal 1982, p. 135.

p. 140 Imitating bound feet: Dickemann 1979, p. 348.

p. 140 Mazarin's wit: Bernier 1987, p. 178, cites Visconti, P., *Mémoires sur la Cour de Louis XIV,* Paris: 1908.

p. 141 The king's bottom button: Orwell 1958, p. 167.

p. 141 Imitating Rothschild trades: Ferguson 1998a, p. 287.

p. 142 "Ambitious subordinates": Boehm 1999, p. 163.

p. 143 Winchilsea and Nottingham: Mosley 1999, vol. 2, p. 3048.

p. 143 Basia's self-interested subordinates: Margolick 1993, pp. 124–26, 610.

Chapter Seven: Why Do Rich People Take Such Risks?

p. 144 Churchill's glowworm remark: Manchester 1983, p. 367.

p. 145 Broad-tailed hummingbirds: Calder and Calder 1992.

p. 145 Bill Gates doesn't give a shit: *Businessweek,* August 14, 2001.

p. 145 Spyder C8 Spyker: *The Wall Street Journal Europe,* April 27, 2001.

p. 146 Pronghorn deer: Byers 1998.

p. 146 Anne and the mandrill's bottom: Botting 1999, p. 375.

p. 146 Sassoon's divine lunch guest: Crook 1999, p. 186.

p. 148 The best discussion of the handicap principle is in Zahavi and Zahavi 1997; bowerbirds appear on p. 22.

p. 149 Packer's $20 million weekend: BBC News, August 31, 2000.

p. 149 Koch's Cowpoke Art: *The Wall Street Journal,* September 7, 2001.

p. 149 Zahavi's maddening contrariness: Dawkins 1989, pp. 159–60.

p. 150 Wild dogs and hyenas: Zahavi and Zahavi 1997, p. 7.

p. 150 Grafen: Dawkins 1989, p. 311–13.

p. 150 "Limitless craziness": Dawkins 1989, p. 313.

p. 150 Sexual selection proposed: Darwin 1871, 1974, pp. 203–46.

p. 151 Perelman's engines: Shah 1996.

p. 154 Pittman's luggage: Krakauer 1997, pp. 117, 168.

Chapter Eight: Inconspicuous Consumption

p. 158 "Conspicuous consumption": Veblen 1899, 1967, p. 84.

p. 158 Gold-plated oysters: Hale 1994, p. 388.

p. 158 Underwater billiards: Crook 1999, p. 157–58.

p. 158 Elvis's extravagant PBJ: *Naples Daily News* (FL), May 17, 1998.

p. 158 Brunei: Richard Behar's excellent report on the kingdom appeared in *Fortune,* February 1, 1999.

p. 158 Beauty expense, expense beauty: Veblen 1899, 1967, p. 132.

p. 158 The silver spoon: Veblen 1899, 1967, pp. 126–28.

p. 159 A stern line against discounting: Catrett and Lynn 1999.

p. 159 "Conspicuous wastefulness": Veblen 1899, 1967, p. 128.

p. 160 "Wonderland of wasteful sexual signaling" and wasteful display in nature: Miller 1999, pp. 21–22.

p. 160 The broad-tailed hummingbird's display: Calder and Calder 1992.

p. 160 Ferrari off the line: Owners tend to be fussy about this sort of thing, so it's actually zero–62 mph/100km in 4.5 seconds.

p. 160 Sennheiser Orpheus Set headphones: Miller 1999, p. 18.

p. 161 Sustainable wastefulness and hand-me-down status: Miller 1999, p. 22.

p. 161 "Conspicuous consumption or conspicuous children": Miller 1999, p. 22.

p. 161 Higher math: Chamberlain seems variously to have boasted of using either ten thousand or seventeen thousand arctic wolf muzzles in the house. The smaller number represents the total population of wolves now thought to be living in Alaska. Chamberlain's claims for success in bed were more modest, not even encompassing all the eligible women then living in California. To arrive at his figure of 20,000 women, he calculated an average of 1.2 women a day from the age of fifteen, including many six-women days, and one Herculean night in which he shared his special magic with fourteen women. Chamberlain made his estimate in 1991 but he lived for another eight years. Even assuming the geriatric rate of only one new woman a day, this would increase his lifetime total to almost 23,000. Chamberlain 1991, pp. 250–68.

p. 162 Chamberlain's house described in *The Los Angeles Times,* March 13, 1972, and July 30, 2001.

p. 162 Getty as Hadrian: Lenzner 1986, p. 108.

p. 162 "The art they admired": Lenzner 1986, p. 178.

p. 162 The Rockefellers' benefactions: Collier and Horowitz 1976, pp. 144–49.

p. 163 Galbraith made the horse and sparrow remark at Tufts University, March 27, 2000.

p. 163 From prestige to practical technologies: Hayden 2001,

p. 163 America's first flush toilet: Goldsmith 1980, p. 136.

p. 163 Balfour's acres in Cannadine 1990, p. 225; his brain and the birth of the country house weekend, Tuchman 1966, pp. 46, 53. The *Oxford English Dictionary* tracks the term back a bit further, to 1878. See also Fussell 1983, p. 107.

p. 164 Freud's VIP rooms: *Vanity Fair,* August 2001, p. 145.

p. 165 Rocancourt in The Hamptons: *New York Magazine,* August 21, 2000; *The New York Times,* October 2, 2000; and *Vanity Fair,* January 2001.

p. 166 Sumptuary laws: Tuchman 1978, p. 19; and Baldwin 1926, p. 53.

p. 166 Badges of status: Roper 1986, pp. 38–40.

p. 166 House sparrows: Zahavi and Zahavi 1997, p. 171; and Møller 1990.

p. 166 Great tits: Zahavi and Zahavi 1997, p. 55.

p. 166 Platyfish tails: Dugatkin 2000, pp. 47–48.

p. 166 European barn swallows: Birkhead 2000, p. 42; and Zahavi and Zahavi 1997, p. 33.

p. 167 General Bob Johnson: Goldsmith 1987, p. 89.

p. 167 The Grand Duke's mazurka: Balsan 1973, pp. 124–25.

p. 168 The fruitfly mazurka: Andersson 1994, p. 73; and Maynard Smith 1956, pp. 261–79.

p. 168 Coburg as a "little ape": Potts and Potts 1995, p. 10.

p. 168 Biological screening in India: Allen 1984, pp. 182–83.

p. 169 Gold, jade, diamonds as badges of status: Clark 1986, pp. 10, 75–76.

p. 170 The Karagwe honeypot wives: Wrangham and Peterson 1996, pp. 162–63.

p. 170 Fat and social status in different cultures: Beller 1977, pp. 261–62.

p. 171 Texas breasts: Stuart-Macadam and Dettwyler 1995, pp. 175–76.

p. 171 The fashion for paleness: Tabori 1961, p. 113.

p. 172 Keeping the common crowd common: Bedford 1966, pp. 18–19.

p. 173 Sumptuary laws sanctifying rank: Sanders 1979, p. 322.

p. 173 "Apparel not pertaining to their estate": Baldwin 1926, p. 46.

p. 173 Clergy opposes extravagance: Tuchman 1978, p. 20.

p. 173 "Penniless grooms": Baldwin 1926, p. 74.

p. 173 Glittering clothes of nobles: Hale 1994, p. 388.

p. 174 Gates's house used 4.7 million gallons of water in 2000: *Eastside Journal* (Bellevue, WA), June 5, 2001.

p. 175 Kress Foundation art and I Tatti gift: Lenzner 1986, p. 180.

p. 175 Signal inflation and satin bowerbirds: Zahavi and Zahavi 1997, p. 59–60.

p. 176 Lace collars like organ pipes: Palliser 1875, 1984, p. 140.

p. 177 "A turkey shaking its feathers": Palliser 1875, 1984, p. 168.

p. 177 The two-foot spoon: Palliser 1875, 1984, p. 140.

p. 177 Queen Elizabeth's starch and ruff-cutting: Palliser 1875, 1984, pp. 311, 313.

p. 177 King Henry III: Palliser 1875, 1984, p. 141.

p. 177 Lord Henry Berkeley: Levey 1983, p. 16.

p. 177 Princess Sophia: Palliser 1875, 1984, p. 320–21.

p. 177 Ben Jonson's joke is from his play *Every Man Out of His Humour.*

p. 177 A signal losing its value: Zahavi and Zahavi 1997, p. 59.

p. 179 Plato in the bathroom was an American Standard ad appearing in 1996.

p. 181 Lady Montdore in her jewels: Mitford 1949, pp. 97–98.

p. 181 Courtesans, déclassé embellishments: Balsan 1973, pp. 48–49.

p. 183 Violent shopping: Wells 1909, p. 287.

p. 183 Alan Grubman's Rolls-Royce: *Talk,* October 2000.

p. 186 Tara Rockefeller: *Vanity Fair,* October 2000.

p. 187 Michael Heseltine: Clark 1993.

p. 187 Mountain gorillas, feral horses, etc.: Watts 1991.

p. 188 "The cash nexus": Aldrich 1988, p. 80.

p. 189 Excluding baser elements: Veblen 1899, 1967, p. 187.

Chapter Nine: Living Large

p. 192 Yellow warblers and chipping sparrows: Cody 1985.

p. 193 The glitter of water: Coss and Moore 1990.

p. 193 Lytton Strachey: Montgomery-Masingberd 1985, p. 12.

p. 194 "The finest view in England": Fowler 1989, p. 164.

p. 194 "Nothing to equal this": Montgomery-Masingberd 1985, p. 73.

p. 194 Walpole and Coward: Montgomery-Masingberd 1985, pp. 11, 182.

p. 194 Voltaire ("A great mass of stone without harmony or taste"): Fowler 1989, p. 74.

p. 194 "The dump": Montgomery-Masingberd 1985, p. 12; "that wild, unmerciful house," Cowles 1983, p. 393.

p. 194 "We shape our dwellings": Fowler 1989, p. 252.

p. 195 Private food storage: Kuijt 2000 and personal communication.

p. 195 Farewell to circular huts: Özdogan 1997, p. 10.

p. 196 Loma Torremote: Wason 1994, p. 142; Sanders et al. 1979.

p. 196 The Big House and its acres; owning 80 percent of the U.K: Cannadine 1990, p. 8.

p. 196 Disraeli: Cannadine 1990, p. 17.

p. 196 The Spellings: *W,* October 2000.

p. 197 Visitors forget their names: Bronfman 1987, p. 109.

p. 197 Gangling Arabella: Cowles 1983, pp. 5, 13.

p. 197 Female coalitions: De Waal 1982, p. 107.

p. 198 "Very little service": Montgomery-Masingberd 1985, p. 21; and Cowles 1983, p. 25.

p. 198 Help from Villiers: Cowles 1983, pp. 39–42.

p. 198 Churchill's prospects: Cowles 1983, p. 26.

p. 198 "The fury heart": Cokayne 1910, p. 496.

p. 199 Anne's physical attraction to Sarah: Cowles 1983, p. 92.

p. 199 The Churchills' income: Cokayne 1910, p. 495; and Cowles 1983, p. 171.

p. 199 "Ever watchful, ever right": Cowles 1983, p. 356.

p. 200 Mayan ancestors: Schele and Freidel 1990, p. 307.

p. 200 Family magic: van Gulik 1961, pp. 12–14.

p. 201 Pomp, putrefaction: Balsan 1973, p. 69.

p. 201 "The Marlboroughs worshipped here": Fowler 1989, p. 221.

p. 201 "The cock Balthazar": Lorenz 1966, p. 31.

p. 202 Cost of building Blenheim: Montgomery-Masingberd 1985, p. 50.

p. 203 *This is all we have*": *W,* October 2000.

p. 204 Vanbrugh and Sarah Churchill on the royal ruins: Montgomery-Masingberd 1985, pp. 44, 60.

p. 204 Blenheim "a chaos": Bond and Tiller 1997, p. 78.

p. 205 Young Churchill at Versailles: Cowles 1983, pp. 34–35.

p. 206 Blenheim's bawdy art: Fowler 1989, pp. 26–29, 157.

p. 207 Prospect-refuge: Appleton 1996, p. 63.

p. 208 Sarah Churchill observing, unobserved: Green 1967, p. 31.

p. 208 Pavlovian experiments: Kellert and Wilson 1993, pp. 77–78.

p. 209 The restorative effects of natural scenes: Kellert and Wilson 1993, pp. 102–6.

p. 209 Appealing elements in a landscape and the Savanna Hypothesis. See Heerwagen and Orians' "Humans, Habitats, and Aesthetics," in Kellert and Wilson 1993, pp. 138–72; and their "Evolved Responses to Landscapes," in Barkow et al. 1992, pp. 555–79.

p. 210 Why we value flowers: Kellert and Wilson 1993, p. 144.

p. 210 Constable's "savannifying": Kellert and Wilson 1993, pp. 155–56.

p. 210 Repton's landscapes: Kellert and Wilson 1993, pp. 154–55.

p. 211 Big scary animals at Blenheim: Bond and Tiller 1997, p. 32.

p. 212 "Dark Deeds at Night": Cowles 1983, p. 295.

p. 213 Rumfoord: Vonnegut describes him in *The Sirens of Titan,* p. 13.

p. 213 Mounting the barricades: Bronfman 1987, p. 117.

p. 214 "It's scary out there": Bronfman 1987, p. 132.

p. 217 "An unpleasant place": Montgomery-Masingberd 1985, p. 13.

p. 217 "Take me away": Montgomery-Masingberd 1985, p. 93.

p. 217 Dissolute dukes: Montgomery-Masingberd 1985, pp. 97–98.

p. 217 Lillian million: Montgomery-Masingberd 1985, p. 111.

p. 218 "Futile expenditure of wealth": Crook 1999, p. 19.

p. 218 Charging admission at Blenheim: Fowler 1989, pp. 112, 147; and Montgomery-Masingberd 1985, p. 12.

p. 218 "Freaks or animals": Montgomery-Masingberd 1985, p. 197.

p. 219 The piggy bank tantrum: Bronfman 1987, pp. 83–84.

p. 219 Growing up in isolation: Bronfman 1987, p. 89.

p. 220 Consuelo Vanderbilt as "a link in the chain" and "that little upstart" Winston: Balsan 1973, pp. 52, 57.

p. 220 Not one livable room: Balsan 1973, p. 65.

p. 220 Dinner with the duke: Balsan 1973, p. 60.

p. 221 Buried at Bladon: Balsan 1973, p. 69.

p. 222 Winston's arrival was impeccably timed in one other regard: The aristocracy was just entering its final decline. New laws were stripping away its land and, without a territorial basis to supply income, big houses everywhere were slipping into ruin. Between 1945 and 1955, four hundred country houses were demolished. (Cannadine 1990, p. 644.) But not Blenheim, which had a new hero to celebrate.

p. 223 "A cannnonball at Malplaquet": Fowler 1989, p. 246.

p. 223 Blenheim fulfilled its purpose: Churchill 1933, vol. 4, p. 319.

Chapter Ten: The Temptation of Midnight Feasts

p. 224 Hangingflies: Thornhill 1980.

p. 225 Donald Trump: *The New York Times,* May 3, 1997; and *People Weekly,* May 19, 1997.

p. 225 Female concern with resources: Social psychologists and anthropologists have documented this attitude in numerous studies, across drastically different cultures. See Perusse 1993, pp. 267, 281; Weiderman and Allgeier 1992, p. 118; and Buss 1994, p. 26.

p. 225 Women with good prospects place greater emphasis on resources: Weiderman and Allgeier 1992, p. 121.

p. 225 "Let's go!": *Forbes,* October 9, 2000.

p. 225 Past survival advantage of male with resources: Buss 1994, p. 25.

p. 226 Common terns: Andersson 1994, p. 187.

p. 226 Sperm counts: Birkhead 2000, pp. 74, 113.

p. 226 Running to Chicago: Burnham and Phelan 2000, p. 142.

p. 228 "I wasn't interested in figures, but . . .": Goldsmith 1987, p. 26.

p. 228 Northern elephant seals: LeBoeuf and Reiter in Clutton-Brock 1988, pp. 344–62; see also Andersson 1994, pp. 118–19, and Birkhead 2000, p. 154.

p. 229 Kinky ducks (black ducks and mallards): Ellis 1995, pp. 273, 283.

p. 229 Shirley does Charlemagne: *New York Post,* May 18, 2000.

p. 230 The *té* of a girl, yin-gathering: Gulik 1961, pp. 13, 26, 46.

p. 230 Positions: Gulik 1961, p. 130.

p. 231 Gnashing of teeth: This is not a sex guide, and you will have to find your own *p'ing-i point.* But Gulik, p. 194, gives useful clues and suggests that a man must imagine his essence splitting into sun and moon, rolling around his abdomen for a bit, then ascending to be reunited in the *Ni-huan* spot in his brain. Happy hunting!

p. 231 Betzig gives her honeypot tallies in Pérusse 1993, p. 285, and by personal communication.

p. 231 Unmarried household staff: Betzig in Smith 2002.

p. 231 Boswell's philandering: Pottle 1953, pp. 247–48.

p. 232 Midcentury census, 1883 investigation, self-help advice, and Jonathan Swift cited by Betzig in Smith 2002.

p. 232 Con Phillips: Stone 1992, pp. 237–74.

p. 232 The duke and the housemaid: Balsan 1973, p. 62.

p. 233 Syphilis blamed on Blenheim maid: Fowler 1989, p. 193.

p. 233 "The polygyny threshold": Verner and Willson 1966 coined the term. See also Orians 1969; and Andersson 1994, pp. 195–99.

p. 234 Enough resources: Smith and Sandell 1998.

p. 234 An eye for real estate: In one study, human females also clearly thought the real estate was important: Among Kipsigis tribespeople on newly settled land in Kenya, researchers found a strong correlation between the size of a man's landholdings and the number of wives. But females apparently suffered a cost: The more wives in a marriage, the fewer children survived per wife. See Andersson 1994, p. 196.

p. 234 Money his only love: Winokur 1996.

p. 234 William Avery Rockefeller behaves badly: Chernow 1998, pp. 8–9, 28; and John D. becomes sweet, respectful husband: Chernow 1998, pp. 121, 124–25.

p. 235 Larry Hillblom: *GQ,* August 1998, p. 214.

p. 235 Edward eyes the audience: Potts and Potts 1995, p. 148.

p. 236 David Lee Roth describes his "take a number" approach to love in his 1997 autobiography *Crazy from the Heat* (New York: Hyperion).

p. 237 Mate-copying in guppies: Dugatkin 2000, pp. 53–85.

p. 237 Sage grouse: Gibson et al. 1991.

p. 237 Human mate-copying: Dugatkin 2000, pp. 75–80.

p. 238 An ugly man with a knockout: Buss 1994, p. 59.

p. 238 Secondary nests at a distance from home: Smith and Sandell 1998.

p. 239 The "charming cohort": Sackville-West 1930, p. 89.

p. 239 J. Paul Getty's bigamy: Lenzner 1986, pp. 30, 42, 44.

p. 239 "A lasting relationship with a woman": *Time,* August 15, 1994.

p. 240 Gordon Getty's happy marriage: Pearson 1995, p. 122.

p. 240 Selling Getty Oil: Pearson 1995, p. 198.

p. 241 Athletic infidelity: "Paternity Ward," *Sports Illustrated,* May 4, 1998.

p. 242 Ecila: Pearsall 1974, p. 30.

p. 242 "Chamberpots": Weintraub 2001, pp. 274–75.

p. 242 Starling extra-pair young: Smith and Sandell 1998.

p. 243 Mr. Average Swallow: Birkhead 2000, p. 227.

p. 243 Macmillan: Horne 1989, pp. 78, 85, 86.

p. 243 Misassigned paternity: Betzig in Perusse 1993, p. 284.

p. 244 Dung beetles: Birkhead 2000, p. 53.

p. 244 Bearded weevil: Gould 1989, pp. 118–19.

p. 244 Copulatory plugs: Eberhard, pp. 142–55.

p. 244 Footbinding for immobilization: Dickemann 1979, p. 348.

p. 244 Clitoridectomy: Hartung 1976, p. 613; and Buss 1994, p. 138–39.

p. 245 Chastity belts: *Tatler,* December 2000.

p. 245 Cattle prods: *The New York Times,* August 7, 1999.

p. 245 Armand Hammer's use of tracking device: Edward J. Epstein in *The New Yorker,* September 23, 1996, p. 40.

p. 245 Servants as mate-guards: Dickemann 1979, p. 337; and Balsan 1973, p. 106.

p. 245 Patricia Duff borrows cell phones: *Vanity Fair,* August, 1999.

p. 246 Wealthier women having fewer sexual partners: Vining 1986, p. 199.

p. 247 Ivana Trump avoiding heterosexual men: Hurt 1993, p. 310.

p. 247 Jimmy Donahue: Wilson 2001, p. 110–11.

p. 247 Evan Frankel: Gaines 1998, pp. 170, 176.

p. 248 Princess Stephanie's disinheritance: *Daily Express,* October 7, 2000.

p. 248 Overly vocal Oor: De Waal 1982, p. 49.

p. 248 Marla's best sex ever: Hurt 1993, pp. 279, 282.

p. 248 Queen Marie Thérèse: Bernier 1987, p. 71.

p. 249 Lillie Langtry: Weintraub 2001, pp. 254, 260.

p. 249 Chickadees: Birkhead 2000, p. 198.

p. 249 Arnold Bennett: Fowler 1989, p. 222.

p. 250 Joyously celebrated affairs: Manchester 1983, p. 90.

p. 250 "The disposition of bedrooms": Sackville-West 1930, p. 17.

p. 250 Lord Wimborne: Asquith 1969, pp. 288–89.

p. 251 "A cabinet reshuffle": Mitford 1949, p. 48.

p. 251 Cousins don't count: Manchester 1983, p. 89.

p. 251 "Animal recklessness": Cecil 1954, p. 6.

p. 252 The Australian superb fairy wren: Birkhead 2000, p. 41.

p. 252 Peniston and William: Cecil 1954, p. 50.

p. 252 Primi Visconti: Bernier 1987, p. 171.

p. 252 Mistress of King George II: Betzig 2002.

p. 253 Lord of the Bedchamber: Cecil 1954, p. 16.

p. 254 "Inverse fertility pattern": Perusse 1993, p. 270.

p. 254 The end of the aristocracy: Cannadine 1990, p. 682; and Lacey 1983, p. 216.

p. 255 Orhan Sadik-Khan: *The New York Daily News,* April 22, 23 and 27, 1998; *The New York Post,* April 22 and 23, 1998.

p. 255 Bedding people of below-stairs class: *Daily Telegraph,* May 31, 1994.

p. 255 Access to birth control: Manchester 1983, p. 89.

p. 255 Contraception, abortion, infanticide in Rome: Brunt 1971, pp. 146–47.

Chapter Eleven: Family Business

p. 257 Codfish: Wodehouse 1922, p. 10.

p. 257 See Lord Bath's Web site: http://www.lordbath.co.uk/.

p. 258 Lady Mordaunt: Weintraub 2001, pp. 164–68.

p. 259 Sir Thomas Thynne: Green 1967, p. 31.

p. 259 A group of five women: *Express News,* December 6, 1999.

p. 260 "Having children": Rubinstein 1980, p. 138.

p. 260 "Shirtsleeves": Ward 1987, p. 1.

p. 260 Rockefeller inheritance: Chernow 1998, pp. 510–12.

p. 261 The dual investment strategy: Kaplan and Hill, in Vining 1986, p. 199.

p. 261 Geffen, Allen, and Mitchell: *Forbes,* October 8, 2001.

p. 261 Reproductive success in the *Forbes* 400: Essock-Vitale 1984.

p. 262 Not Oedipus: Hartung 1982, pp. 1–12.

p. 263 Predestined wild oats: Sackville-West 1930, p. 63.

p. 263 Malcolm Forbes: Nass 1991, p. 261.

p. 263 Sex ratio in presidential children: See Ridley 1993, p. 115, for a discussion of Trivers-Willard. See Trivers and Willard 1973.

p. 264 The tanner's daughter: Mosley 1999, p. xxxi.

p. 265 Posthumous mate-guarding: Betzig 1992c, p. 60; and Betzig 1992b, p. 370.

p. 265 Henry Ford II: Winokur 1996, p. 20.

p. 266 Attractions to kin and high rank: Cheney and Seyfarth 1990, p. 180.

p. 266 "Lucky sperm club": *The Wall Street Journal,* March 10, 2000; and Fred Trump's estate: Hurt 1993, p. 424.

p. 266 *Bright Young Things: Vanity Fair,* October 2000.

p. 266 Rockefeller, Jr., envies the office girls: Chernow 1998, p. 510.

p. 266 Richard Li: *Forbes,* July 5, 1999; *The Washington Post,* March 2, 2000; *Fortune,* March 19, 2001.

p. 267 "A great big person standing behind me": Bronfman 1987, p. 80.

p. 268 "The glue of generations": Birmingham 1987, p. 76.

p. 268 Du Pont inbreeding: Mosley 1980, p. 38; Thorndike 1976, p. 268; and Allen 1989, p. 115.

p. 269 Rothschild inbreeding: Ferguson 1998a, pp. 74, 188–89; Ferguson 1998b, pp. xxvi, 11–13, 242.

p. 270 Royal incest: Berghe and Mesher 1980.

p. 270 The first surviving daughter in two hundred years: Allen 1984, p. 25.

p. 270 European gentry also got rid of excess daughters: Dickemann 1979, p. 321; and Betzig 1992b, p. 358.

p. 270 Male-female bias in nursing: Gaulin and Robbins 1991.

p. 271 Genealogy of Prince Charles: Shoumatoff 1985, p. 234; and Gunderson, personal communication.

p. 271 Four Rothschild grandsons: Ferguson 1998a, pp. 184–85.

p. 271 The inbreeding coefficient analysis was kindly provided by Mark Urban at Yale University School of Forestry. Urban writes: "Determination of a 'normal' genetic relatedness depends on your assumptions of the effective size of the mating population for the family. For example, assuming that family members could potentially marry the 10,000 people that surrounded them, then the norm for five generations would be an inbreeding coefficient of 0.00025. . . . Re-writing the formula and solving for population size, we find that the children of Albert and Bettina have the same inbreeding coefficient as if the previous five generations of their family had been living on an island comprised of 38 individuals."

p. 272 The tyrannical mouse model is discussed in Crowcroft, Peter, 1966. *Mice All Over,* London: G. T. Foulis.

p. 272 A "gene for financial acumen": Ferguson 1998a, p. 189.

p. 272 Insects named for Rothschilds: Ferguson 1998a, p. 1.

p. 273 "Families of cousins": Aldrich 1988, p. 55.

p. 274 Trying to smell how other children were different: Bronfman 1987, p. 73.

p. 275 Owning 4 million acres: Winchester 1982, p. 211.

p. 275 Ruling 400 million people: Fowler 1989, p. 180; and one-quarter of the Earth: Tuchman 1966, pp. 54–55.

p. 275 Aristocrats in the 1999 Rich List: *Economist,* June 5, 1999.

p. 275 American daughters to titled Europeans: Jaher 1980, p. 200.

p. 275 Jamie Lee Curtis: *People*, February 8, 1999.

p. 276 "The three or four hundred families": Fowler 1989, p. 227.

p. 276 "A collective awareness": Cannadine 1990, p. 24.

p. 276 The Cecil kin: Cannadine 1990, p. 203.

p. 277 The cousinship: Cannadine 1990, pp. 188, 195.

p. 277 Dipping into the green: Winchester 1982, p. 249.

p. 279 Honeymoon reading: Balsan 1973, p. 45.

p. 280 Aristocratic world like Chinese grammar: Cannadine 1990, p. 16.

p. 280 "Impossible acts": Freeman, E. A. 1878. "Pedigrees and Pedigree Makers." *Contemporary Review* XXX: 11–41.

p. 280 "The best thing in fiction": Wilde's *A Woman of No Importance.*

p. 281 "Perhapsburgs": Round 1901, pp. 216–49.

p. 281 Undistinguished Churchills: Round 1930, pp. 33–42.

p. 282 "All civilization has been the work of aristocracies": Cannadine 1990, p. 50.

p. 282 "Impeccable connections": Cannadine 1990, p. 220.

p. 282 Kennedy and Bush dynasties: *The New York Times,* January 11, 2001.

p. 283 Ancestral demand on my womb: Bronfman 1987, p. 357.

p. 285 "A horrible instrument": Balsan 1973, pp. 10–11.

p. 285 "Necessary ornaments": Bronfman 1987, p. 30.

p. 285 J. Paul Getty's son and dog: Lenzner 1986, p. 112–13.

p. 286 *"A finished specimen"*: Balsan 1973, pp. 10–11. Italics added.

p. 286 Too many Nordstroms: *The Wall Street Journal,* September 8, 2000.

p. 287 "Inheritance powder": Bernier 1987, pp. 202–3.

p. 287 The "don't-murder-me" plan: Goldsmith 1988, p. 266.

p. 287 Hoogstraten: *The Observer Magazine,* November 8, 1987.

p. 288 Courtney Sale Ross in eternity: *W,* September 2000.

p. 288 Josie Arlington: Marion 1977, pp. 103–104.

p. 288 Doris Duke as "selfish, self-centered": Duke 1996, p. 264.

p. 288 Doris Duke among the sharks: Duke 1996, p. 225.

Epilogue: A How-To Guide for Alpha Apes

p. 291 Women wanting sperm: reported in the *Hartford Courant* (CT), February 14, 2000, by Mary Jo Kochakian, who comments, "Well, is that ever wrong."

p. 291 The monkey's behind: De Waal 1996, p. 97, attributes it to St. Bonaventura, but it has also been attributed elsewhere to General Joseph Stilwell.

p. 291 "Anal types": Jaher 1980, p. 190.

p. 291 "Cupidity and collecting mania": Otto Fenichel quoted in Muensterberger 1994, p. 21.

p. 292 The "abnormal and scandalous" Freud and his "erotic equivalents": Elsner and Cardinal 1994, pp. 229, 233.

p. 294 Irl Engelhardt: Jeff Goodell reporting in the *New York Times Magazine*, July 22, 2001.

p. 294 Conditioning for success: Weisfeld 1980, p. 275.

p. 294 Chinese crickets: Hofmann 2000.

p. 294 Rockefeller's drag racing: Chernow 1998, p. 120.

p. 295 "Ethologically despotic versus culturally egalitarian": Boehm 1999, p. 173.

p. 295 Wishful thinking: Boehm 1999, p. 10.

p. 295 The egalitarianism of Louis XIV: Bernier 1987, p. 97.

p. 296 Getty and his yachts: Lenzner 1986, p. 45.

p. 297 "His-and-her escape passages": *W,* September 2000.

Bibliography

Adler, N. E., E. S. Epel, et al. 2000. "Relationship of Subjective and Objective Social Status with Psychological and Physiological Function: Preliminary Data in Healthy White Women." *Health Psychology* 19, no. 6: 544–50.

Aldrich, N. W., Jr. 1988. *Old Money: The Mythology of America's Upper Class*. New York: Knopf.

Allen, C. 1975. *Plain Tales from the Raj*. London: Abacus.

Allen, C. 1984. *Lives of the Indian Princes*. London: Century Publishing.

Allen, P. M. 1989. *The Founding Fortunes—A New Anatomy of the Super-Rich Families in America*. New York: E.P. Dutton.

Altmann, S. A. 1967. *Social Communication among Primates*. Chicago: University of Chicago Press.

Andersen, K. 1999. *Turn of the Century: A Novel*. New York: Random House.

Andersson, M. 1994. *Sexual Selection*. Princeton: Princeton University Press.

Appleton, J. 1976. *The Experience of Landscape*. New York: Wiley.

Asquith, C. 1969. *Diaries 1915–1918*. New York: Knopf.

Auletta, K. 1997. *The Highwaymen: Warriors of the Information Superhighway*. New York: Random House.

Baldwin, F. E. 1926. *Sumptuary Legislation and Personal Regulation in England*. Baltimore: Johns Hopkins Press.

Balsan, C. V. 1973. *The Glitter and the Gold*. Maidstone: George Mann.

Balzac, H. 1835, 1954. *Pere Goriot*. New York: Dodd, Mead.

Barkow, J. 1989. *Darwin, Sex, and Status: Biological Approaches to Mind and Culture*. Toronto: University of Toronto Press.

Barkow, J. H., L. Cosmides, and J. Tooby. 1992. *The Adapted Mind: Evolutionary Psychology and the Generation of Culture*. New York: Oxford University Press.

Bedford, J. 1966. *The Book of Snobs*. New York: Coward-McCann.

Beller, A. S. 1977. *Fat and Thin: A Natural History of Obesity*. New York: Farrar, Straus.

Bender, B. 1978. "Gatherer-hunter to Farmer: A Social Perspective." *World Archaeology* 10, no. 2, pp. 204–22.

Berghe, P. L. van den, and G. M. Mesher. 1980. "Royal Incest and Inclusive Fitness." *American Ethnologist* 7: 300–17.

Bernier, O. 1987. *Louis XIV: A Royal Life.* New York: Doubleday.

Bernstein, I. S. 1981. "Dominance: The Baby and the Bathwater." *The Behavioral and Brain Sciences* 4: 419–57.

Betzig, L. 1992a. "Roman Polygyny." *Ethology and Sociobiology* 13: 309–49.

Betzig, L. 1992b. "Roman Monogamy." *Ethology and Sociobiology* 13: 351–83.

Betzig, L. 1992c. "Sex, Succession and Stratification in the First Six Civilizations." In *Socioeconomic Inequality and Social Stratification,* ed. L. Ellis. New York: Praeger.

Betzig, L. 2002. "British Polygyny." In *Human Biology and History,* ed. M. Smith. London: Taylor & Francis.

Bibb, P. 1993. *It Ain't As Easy As It Looks.* New York: Crown.

Bingham, S. 1989. *Passion and Prejudice: A Family Memoir.* New York: Knopf.

Birkhead, T. 2000. *Promiscuity: An Evolutionary History of Sperm Competition.* Cambridge: Harvard University Press.

Birmingham, S. 1967. *Our Crowd: The Great Jewish Families of New York.* New York: Harper & Row.

Birmingham, S. 1987. *America's Secret Aristocracy.* Boston, MA: Little Brown & Co.

Boehm, C. 1999. *Hierarchy in the Forest: The Evolution of Egalitarian Behaviors.* Cambridge: Harvard University Press.

Bond, J., and K. Tiller. 1997. *Blenheim: Landscape for a Palace.* Stroud: Sutton.

Boone, J. L., and K. L. Kessler. 1999. "More Status or More Children? Social Status, Fertility Reduction and Long-Term Fitness." *Evolution and Human Behavior* 20: 257–77.

Borgerhoff Mulder, M. 1990. "Kipsigis Women's Preferences for Wealthy Men: Evidence for Female Choice in Mammals." *Behavioural Ecology and Sociobiology* 27: 255–64.

Botting, D. 1999. *Gerald Durrell: The Authorized Biography.* New York: Carroll and Graf.

Branson, R. 1998. *Losing My Virginity: How I've Survived, Had Fun, and Made a Fortune Doing Business My Way.* New York: Times Books.

Bronfman, J. 1987. *The Experience of Inherited Wealth: A Social-Psychological Perspective.* Ann Arbor, MI: ProQuest.

Bruck, C. 1988. *The Predators' Ball.* New York: Simon & Schuster.

Brunt, P. A. 1971. *Italian Manpower: 225 B.C.–A.D. 14.* Oxford: Clarendon.

Burnham, T., and J. Phelan. 2000. *Mean Genes: From Sex to Money to Food, Taming our Primal Instincts.* New York: Penguin.

Buss, D. M. 1994. *The Evolution of Desire.* New York: BasicBooks.

Busvine, J. R. 1976. *Insects, Hygiene and History.* London: University of London Press.

Bibliography

Byers, J. A. 1998. *American Pronghorn: Social Adaptations & The Ghosts of Predators Past.* Chicago: University of Chicago Press.

Calder, W. A., and L. L. Calder. 1992. "The Broad-tailed Hummingbird." Number 16 in *Birds of North America,* eds. A. Poole and P. Stettenheim. Philadelphia, PA: American Ornithologists' Union and Academy of Natural Sciences.

Cannadine, D. 1990. *The Decline and Fall of the British Aristocracy.* New Haven, CT: Yale University Press.

Carroll, D., G. D. Smith, and P. Bennett. 1996. "Some Observations on Health and Socioeconomic Status." *Journal of Health Psychology* 1, no. 1: 23–39.

Catrett, J., and M. Lynn. 1999. "Managing Status in the Hotel Industry: How Four Seasons Comes to the Fore." *Cornell Hotel & Restaurant Administration Quarterly* 40.

Cecil, D. 1954. *Melbourne.* London: The Reprint Society.

Chamberlain, W. 1991. *A View from Above.* New York: Villard.

Cheney, D., and R. Seyfarth. 1990. *How Monkeys See the World: Inside the Mind of Another Species.* Chicago: University of Chicago Press.

Chernow, R. 1998. *Titan: The Life of John D. Rockefeller, Sr.* New York: Random House.

Churchill, W. S. 1933. *Marlborough: His Life and Times.* London: Harrap.

Clark, A. 1993. *Diaries of Alan Clark.* London: Weidenfeld.

Clark, G. 1986. *Symbols of Excellence—Precious Materials as Expressions of Status.* Cambridge: Cambridge University Press.

Clutton-Brock, T. H. (ed.) 1988. *Reproductive Success.* Chicago: University of Chicago Press.

Cody, M. 1985. *Habitat Selection in Birds.* New York: Academic Press.

Cokayne, G. E. 1910. *The Complete Peerage of England, Scotland, Ireland, Great Britain and the United Kingdom: Extant, Extinct, or Dormant.* London: The St. Catherine Press.

Collier, P., and D. Horowitz. 1976. *The Rockefellers: An American Dynasty.* New York: Holt, Reinhart, and Winston.

Coss, R. G., and M. Moore. 1990. "All That Glistens: Water Connotations in Surface Finishes." *Ecological Psychology* 2, no. 4: 367–80.

Coulanges, F. de. 1956. *The Ancient City: A Study on the Religion, Laws and Institutions of Greece and Rome.* Garden City, NY: Doubleday.

Cowles, V. 1983. *The Great Marlborough and His Duchess.* New York: Macmillan.

Crook, J. M. 1999. *The Rise of the Nouveaux Riches: Style and Status in Victorian and Edwardian Architecture.* London: John Murray.

Darwin, C. 1859, 1968. *The Origin of Species by Means of Natural Selection.* New York: Penguin Books.

Darwin, C. 1871, 1974. *The Descent of Man and Selection in Relation to Sex.* Detroit: Gale Research.

Dawkins, R. 1989. *The Selfish Gene.* Oxford: Oxford University Press.

Dawkins, R. 1994. "Gaps in the Mind." In *The Great Ape Project: Equality beyond Humanity,* eds. P. Cavalieri and P. Singer. New York: St. Martin's.

De Waal, F. 1982. *Chimpanzee Politics: Power and Sex among Apes.* Baltimore: Johns Hopkins University Press.

De Waal, F. 1996. *Good Natured: The Origins of Right and Wrong in Humans and Other Animals.* Cambridge: Harvard University Press.

De Waal, F. 1997. *Bonobo: The Forgotten Ape.* Berkeley, CA: University of California Press.

Diamond, J. 1992. *The Third Chimpanzee: The Evolution and Future of the Human Animal.* New York: HarperCollins.

Dickemann, M. 1979. "Female Infanticide, Reproductive Strategies, and Social Stratification: A Preliminary Model." In *Evolutionary Biology and Human Social Behavior: An Anthropological Perspective,* eds. N. A. Chagnon and W. Irons. North Scituate, MA: Duxbury Press.

Dixson, A. F. 1998. *Primate Sexuality—Comparative Studies of the Prosimians, Monkeys, Apes, and Human Beings.* Oxford: Oxford University Press.

Dugatkin, L. A. 2000. *The Imitation Factor: Evolution beyond the Gene.* New York: The Free Press.

Duke, P., and J. Thomas. 1996. *Too Rich: The Family Secrets of Doris Duke.* New York: HarperCollins.

Dunbar, R. 1996. *Grooming, Gossip, and the Evolution of Language.* Cambridge: Harvard University Press.

Eberhard, W. G. 1996. *Female Control: Sexual Selection by Cryptic Female Choice.* Princeton: Princeton University Press.

Eibl-Eibesfeldt, I. 1972. *Love and Hate: The Natural History of Behavior Patterns.* Trans. Geoffrey Strachan. New York: Holt, Rinehart, and Winston.

Ellis, L. 1995. "Dominance and Reproductive Success among Nonhuman Animals: A Cross-Species Comparison." *Ethology and Sociobiology* 16:257–333.

Elsner, J., and R. Cardinal (eds.). 1994. *The Cultures of Collecting.* Cambridge: Harvard University Press.

Erikson, E. H. 1966. "Ontogeny of Ritualisation in Man," *Philosophical Transactions of the Royal Society of London Series B, Biological Sciences* 251: 337–49.

Essock-Vitale, S. 1984. The Reproductive Success of Wealthy Americans. *Ethology and Sociobiology* 5: 45–49.

Etcoff, N. 1999. *Survival of the Prettiest: The Science of Beauty.* New York: Doubleday.

Ferguson, N. 1998a. *The House of Rothschild: Money's Prophets: 1798–1848.* New York: Penguin.

Ferguson, N. 1998b. *The House of Rothschild: The World's Banker: 1849–1999.* New York: Penguin.

Forrest, D. 1970. *Tiger of Mysore: The Life and Death of Tipu Sultan.* London: Chatto & Windus.

Fowler, M. 1989. *Blenheim: Biography of a Palace.* New York: Viking.

Fussell, P. 1983. *Class: A Guide Through the American Status System.* New York: Summit.

Gaines, S. 1998. *Philistines at the Hedgerow: Passion and Property in the Hamptons.* Boston, MA: Little Brown.

Gathorne-Hardy, J. 1973. *The Rise and Fall of the British Nanny.* New York: Dial.

Gaulin, S. J., and C. Robbins. 1991. "Trivers-Willard Effect in Contemporary North American Society." *American Journal of Physical Anthropology* 85:61–69.

Gibson, R. M., J. W. Bradbury, and S. L. Vehrencamp. 1991. "Mate Choice in Lekking Sage Grouse Revisited: The Roles of Vocal Display, Female Site Fidelity and Copying." *Behavioral Ecology* 2: 165–80.

Goldberg, R., and G. J. Goldberg. 1995. *Citizen Turner: The Wild Ride of an American Tycoon.* New York: Harcourt.

Goldblatt, P. (ed.) 1990. *Longitudinal Study: Mortality and Social Organization.* London: HMSO.

Goldsmith, B. 1980. *Little Gloria . . . Happy At Last.* New York: Dell.

Goldsmith, B. 1987. *Johnson v. Johnson.* New York: Knopf.

Goodall, J. 1986. *The Chimpanzees of Gombe: Patterns of Behavior.* Cambridge: Harvard University Press.

Gould, J. 1989. *Sexual Selection.* New York: Scientific American Library.

Gowaty, P. A., and W. C. Bridges. 1991. "Behavioral, Demographic, and Environmental Correlates of Uncertain Parentage in Eastern Bluebirds." *Behavioral Ecology* 2: 339–50.

Green, D. 1967. *Sarah Duchess of Marlborough.* New York: Scribner's.

Gregory, S. W., and S. Webster. 1996. "A Nonverbal Signal in Voices of Interview Partners Effectively Predicts Communication Accommodation and Social Status Perceptions." *Journal of Personality and Social Psychology* 70, no. 6: 1231–40.

Griffin, N., and K. Masters. 1996. *Hit & Run: How Jon Peters and Peter Guber Took Sony for a Ride in Hollywood.* New York: Simon & Schuster.

Gulik, R. H. van. 1961. *Sexual Life in Ancient China.* Leiden: Brill.

Hale, John. 1994. *The Civilization of Europe in the Renaissance.* New York: Atheneum.

Hartung, J. 1976. "On Natural Selection and the Inheritance of Wealth." *Current Anthropology* 17, no. 4: 607–22.

Hartung, J. 1982. "Polygyny and Inheritance of Wealth." *Current Anthropology* 23, no. 1 (February): 1–12.

Hayden, B. 1990. "Nimrods, Piscators, Pluckers, and Planters: The Emergence of Food Production." *Journal of Anthropological Archaeology* 9: 31–69.

Hayden, B. 1998. "Practical and Prestige Technologies." *Journal of Archaeological Methods and Theory* 5: 1–55.

Hayden, B. 2001. "The Origins of Social and Economic Inequality." In *Archaeology in the Millennium*, eds. G. Feinman and D. Price. New York: Plenum.

Hawley, P., and T. D. Little. 1999. "On Winning Some and Losing Some: A Social

Relations Approach to Social Dominance in Toddlers." *Merrill-Palmer Quarterly* 45, no. 2: 185–225.

Hofmann, H., and P. A. Stevenson. 2000. "Flight Restores Fight in Crickets." *Nature* 403: 613.

Hoffmann, H., et al. 1999. "Social Status Regulates Growth Rate: Consequences for Life-History Strategies." *Proceedings of the National Academy of Sciences* 96, no. 24: 14171–6.

Holbrook, S. H. 1953. *The Age of the Moguls.* New York: Doubleday.

Horne, A. 1989. *Harold Macmillan.* New York: Viking.

Hurt, H. 1993. *Lost Tycoon: The Many Lives of Donald J. Trump.* New York: W. W. Norton and Company.

Jaher, F. C. (ed.) 1973. *The Rich, the Well Born, and the Powerful: Elites and Upper Classes in History.* Urbana, IL: University of Illinois.

Jaher, F. C. 1980. "The Gilded Elite: American Multimillionaires, 1865 to the Present." In *Wealth and the Wealthy in the Modern World,* ed. W. D. Rubenstein. London: Croom Helm Ltd.

Kellert, S. R. and E. O. Wilson (eds.). 1993. *The Biophilia Hypothesis.* Washington, DC: Island Press.

Kessler, R. 1999. *The Season: Inside Palm Beach and America's Richest Society.* New York: HarperCollins.

Krakauer, J. 1997. *Into Thin Air.* New York: Villard.

Kuijt, I. 2000. *Life in Neolithic Farming Communities: Social Organization, Identity, and Differentiation.* New York: Kluwer Academic.

Lacey, R. 1983. *Aristocrats.* Boston, MA: Little Brown.

Lenzner, R. 1986. *The Great Getty: The Life and Loves of J. Paul Getty—Richest Man in the World.* New York: Crown Publishers.

Levey, S. M. 1983. *Lace: A History.* London: W. S. Maney & Son.

Lev-Yadun, S., A. Gopher, et al. 2000. "The Cradle of Agriculture." *Science 2* 288 (June): 1602.

Lewis, M. 1989. *Liar's Poker: Rising Through the Wreckage on Wall Street.* New York: W. W. Norton and Company.

Lewis, M. 2000. *The New New Thing.* New York: W. W. Norton and Company.

Lord, W. 1955. *A Night to Remember.* New York: Holt.

Lorenz, K. 1966. *On Aggression.* New York: Harcourt.

Mace, R. 2000. "Evolutionary Ecology of Human Life History," *Animal Behaviour* 59: 1–10.

Manchester, W. R. 1983. *The Last Lion: Winston Spencer Churchill.* Boston, MA: Little, Brown.

Manes, S., and P. Andrews. 1993. *Gates.* New York: Touchstone.

Margolick, D. 1993. *Undue Influence: The Epic Battle for the Johnson & Johnson Fortune.* New York: William Morrow.

Marion, J. F. 1977. *Famous and Curious Cemeteries.* New York: Crown.

Masters, R. D., and M. T. McGuire (eds.). 1994. *The Neurotransmitter Revolution:*

Serotonin, Social Behavior, and the Law. Carbondale, IL: Southern Illinois University Press.

Masters, K. 2000. *The Keys to the Kingdom: How Michael Eisner Lost His Grip.* New York: Morrow.

Maynard Smith, J. 1956. "Fertility, Mating Behavior and Sexual Selection in Drosophila Subobscura." *Journal of Genetics* 54: 261–79.

Mazur, A., and A. Booth. 1998. "Testosterone and Dominance in Men," *Behavioral and Brain Sciences* 21, no. 3: 353–97.

Miller, G. 1999. "Waste Is Good." *Prospect* (February): 18–23.

Miller, G. F. 2000. *The Mating Mind: How Sexual Choice Shaped the Evolution of Human Nature.* New York: Doubleday.

Mitchell, R. 1988. *Civil War Soldiers.* New York: Viking.

Mitford, N. 1949. *Love in a Cold Climate.* New York: Random House.

Møller, A. P. 1990. "Sexual Behavior Is Related to Badge Size in the House Sparrow *Passer Domesticus.*" *Behavioural Ecology and Sociobiology* 27: 23–29.

Montgomery-Masingberd, H. 1985. *Blenheim Revisited.* New York: Beaufort Books.

Morris, D. 1969. *The Human Zoo.* New York: Dell.

Morton, F. 1962. *The Rothschilds: A Family Portrait.* New York: Atheneum.

Mosley, C. 1999. *Burke's Peerage and Baronetage.* London: Fitzroy Dearborn.

Mosley, L. 1980. *Blood Relations: The Rise and Fall of the du Ponts of Delaware.* New York: Atheneum.

Muensterberger, W. 1994. *Collecting: An Unruly Passion.* Princeton: Princeton University Press.

Nass, H. E. 1991. *Wills of the Rich & Famous.* New York: Warner.

Orians, G. 1969. "On The Evolution of Mating Systems in Birds and Mammals." *The American Naturalist* 103, no. 934.

Orwell, G. 1958. *The Road to Wigan Pier.* New York: Harcourt Brace.

Özdogan, M. 1997. "The Beginning of Neolithic Economies in Southeastern Europe: An Anatolian Perspective." *Journal of European Archaeology* 5, no. 2:1–33.

Palliser, B. 1875, 1984. *History of Lace.* New York: Dover.

Pearsall, R. 1974. *Edwardian Life and Leisure.* New York: St. Martin's.

Pearson, J. 1995. *Painfully Rich: The Outrageous Fortune and Misfortunes of the Heirs of J. Paul Getty.* New York: St. Martin's.

Perusse, D. 1993. "Cultural and Reproductive Success in Industrial Societies: Testing the Relationship at the Proximate and Ultimate Levels." *Behavioral and Brain Sciences* 16: 267–322.

Pizzari, T., and T. R. Birkhead. 2000. "Female Feral Fowl Eject Sperm of Subdominant Males." *Nature* 15 (June).

Pottle, F. A. (ed.) 1953. *Boswell on the Grand Tour: Germany and Switzerland 1764.* London: Heinemann.

Potts, D. M., and W.T.W. Potts, 1995. *Queen Victoria's Gene.* Gloucestershire: Alan Sutton Publishing.

Price, J. 1995. "A Remembrance of Thorleif Schjelderup-Ebbe." *Human Ethology Bulletin* 10, no. 1:1–7.

Raleigh, M. J., M. T. McGuire, et al. 1984 "Social and Environmental Influences on Blood Serotonin Concentrations in Monkeys." *Archives of General Psychiatry* 41: 405–10.

Rasmussen, L.E.I., T. D. Lee, et al. 1996. "Insect Pheromones in Elephants." *Nature* 379, no. 6567: 684.

Ridley, M. 1993. *The Red Queen: Sex and the Evolution of Human Nature.* New York: Penguin.

Rivlin, G. 1999. *The Plot to Get Bill Gates: An Irreverent Investigation of the World's Richest Man—And the People Who Hate Him.* New York: Times Business.

Roper, T. 1986. *"Badges of Status in Avian Societies." New Science* 109: 38–40.

Round, J. H. 1901. *Studies in Peerage and Family History.* London: Archibald Constable.

Round, J. H. 1930. *Family Origins and Other Studies.* London: Constable.

Rubinstein, W. D. 1980. *Wealth and the Wealthy in the Modern World.* London: Croom Helm Ltd.

Sackville-West, V. 1930. *The Edwardians.* New York: Doubleday.

Sanders, W. T., et al. 1979. *The Basin of Mexico: Ecological Processes in the Evolution of a Civilization.* New York: Academic Press.

Sapir, E. 1949. "The Unconscious Patterning of Behavior in Society." In *Selected Writings of Edward Sapir in Language Culture and Personality,* ed. D. G. Mandelbaum. Berkeley, CA: University of California Press.

Schaal, B., et al. 1996. "Male Testosterone Linked to High Social Dominance but Low Physical Aggression in Early Adolescence." *Journal of American Academy of Child Adolescent Psychiatry* 35, no. 10: 1322–30.

Schele, L., and D. Freidel. 1990. *A Forest of Kings: The Untold Story of the Ancient Maya.* New York: Morrow.

Schjelderup-Ebbe, T. 1935. "Social Behaviour of Birds." In *Handbook of Social Psychology,* ed. C. Muirchison. Worcester, MA: Clarke University Press, pp. 947–72.

Schumacher, A. 1982. "On the Significance of Stature in Human Society." *Journal of Human Evolution* 11: 697–701.

Shah, D. K. 1996. "Beauty and the Billionaire." *Los Angeles Magazine* 41, no. 12: 62–69.

Shoumatoff, A. 1985. *The Mountain of Names: A History of the Human Family.* New York: Vintage.

Smith, H. G., and M. I. Sandell. 1998. "Intersexual Competition in a Polygynous Mating System," *Oikos* 83: 484–95.

Stanley, T. J., and W. D. Danko. 1996. *The Millionaire Next Door.* Atlanta, GA: Longstreet Press.

Steichen, E. 1963. *A Life in Photography.* New York: Doubleday.

Stengel, R. 2000. *You're Too Kind: A Brief History of Flattery.* New York: Simon & Schuster.

Stone, L. 1992. *Uncertain Unions: Marriage in England 1660-1753*. Oxford: Oxford University Press.

Stove, D. C. 1995. *Darwinian Fairytales*. Aldershot, England: Avebury.

Strassman, J. E. et al. 2000. "Altruism and social cheating in the social amoeba *Dictyostelium discoideum*." *Nature* 408: 965–67.

Strouse, J. 1999. *Morgan: American Financier*. New York: Random House.

Stuart-Macadam, P., and K. A. Dettwyler. 1995. *Breastfeeding: Biocultural Perspectives*. New York: Aldine De Gruyter.

Sykes, C. S. 1982. *Black Sheep*. New York: Viking Press.

Tabori, P. 1961. *The Art of Folly*. Philadelphia: Chilton.

Thorndike, J. J. 1976. *The Very Rich: A History of Wealth*. New York: American Heritage.

Thornhill, R. 1980. "Mate Choice in *Hylobittacus Apicalis* (Insecta: Mecoptera) and Its Relation to Some Models of Female Choice." *Evolution* 34: 519–38.

Trexler, R. C. 1973. "Infanticide in Florence: New Sources and First Results." *History of Childhood Quarterly* 1, no. 1: 98–116.

Trivers, R. L., and D. E. Willard. 1973. "Natural Selection of Parental Ability to Vary the Sex Ratio of Offspring." *Science* 179 (January 5): 90–92.

Tuchman, B. W. 1966. *The Proud Tower*. New York: Macmillan.

Tuchman, B. W. 1978. *A Distant Mirror: The Calamitous 14th Century*. New York: Knopf.

Vander Wall, S. B. 1990. *Hoarding in Animals*. Chicago: University of Chicago Press.

Vanderbilt, A. T. 1989. *Fortune's Children: The Fall of the House of Vanderbilt*. New York: William Morrow.

Veblen, T. 1899, 1967. *The Theory of the Leisure Class*. New York: Viking Penguin.

Verner, J., and M. F. Willson. 1966. "The Influence of Habitats on Mating Systems of North American Passerine Birds." *Ecology* 47, no. 1: 143–7.

Vining, D. R. 1986. "Social versus Reproductive Success: The Central Theoretical Problem of Human Sociobiology." *Behavioral and Brain Sciences* 9: 167–216.

Virgin, C. E., and R. M. Sapolsky. 1997. "Styles of Male Social Behavior and Their Endocrine Correlates among Low-Ranking Baboons." *American Journal of Primatology* 42: 25–39.

Ward, G. 1983. *The Maharajas*. Chicago: Stonehenge.

Ward, J. L. 1987. *Keeping the Family Business Healthy: How to Plan for Continuing Growth, Profitability, and Family Leadership*. San Francisco: Jossey-Bass.

Wason, P. K. 1994. *The Archaeology of Rank*. Cambridge: Cambridge University Press.

Watts, D. P. 1991. "Harassment of Immigrant Female Mountain Gorillas by Resident Females." *Ethology* 89: 135–53.

Weintraub, S. 2001. *Edward the Caresser: The Playboy Prince Who Became Edward VII*. New York: The Free Press.

Weisfeld, G. E. 1980. "Social Dominance and Human Motivation." In *Dominance Relations: An Ethological View of Human Conflict and Social Interaction*, eds. D. R. Omark et al. New York: Garland STPM.

Weisfeld, G. E., and J. M. Beresford. 1982. "Erectness of Posture as an Indicator of Dominance or Success in Humans." *Motivation and Emotion* 6, no. 2: 113–31.

Wells, H. G. 1909, 1931. *Tono-Bungay.* New York: Modern Library.

Wiederman, M. W., and E. R. Allgeier. 1992. "Gender Differences in Mate Selection Criteria: Sociobiological or Socioeconomic Explanation?" *Ethology and Sociobiology.* 13: 115–24.

Wilson, C. 2001. *Dancing with the Devil: The Windsors and Jimmy Donahue.* New York: HarperCollins.

Wilson, M. 1997. *The Difference between God and Larry Ellison: Inside Oracle Corporation.* New York: William Morrow.

Winchester, S. 1982. *Their Noble Lordships: Class and Power in Modern Britain.* New York: Random House.

Winokur, J. 1996. *The Rich Are Different.* New York: Pantheon.

Wodehouse, P. G. 1922. *Blandings Castle.* London: Jenkins.

Wrangham, R., and D. Peterson. 1996. *Demonic Males: Apes and the Origins of Human Violence.* Boston: Houghton Mifflin.

Wright, R. 1994. *The Moral Animal: Evolutionary Psychology and Everyday Life.* New York: Pantheon.

Zahavi, A., and A. Zahavi. 1997. *The Handicap Principle: A Missing Piece of Darwin's Puzzle.* Oxford: Oxford University Press.

Zivin, G. 1977. "Preschool Childrens' Facial Gestures Predict Conflict Outcomes." *Social Science Information* 16, no. 6: 715–30.

PERMISSION CREDITS

Grateful acknowledgment is made to the following for permission to reprint copyrighted materials:

Index